CHROMATOGRAPHY AND SEPARATION SCIENCE

This is Volume 4 of
SEPARATION SCIENCE AND TECHNOLOGY
A reference series edited by Satinder Ahuja

CHROMATOGRAPHY AND SEPARATION SCIENCE

Satinder Ahuja
Ahuja Consulting
Calabash, North Carolina

ACADEMIC PRESS
An imprint of Elsevier Science

Amsterdam   Boston   London   New York   Oxford   Paris   San Diego
San Francisco   Singapore   Sydney   Tokyo

This book is printed on acid-free paper. ∞

Copyright © 2003 Elsevier Science (USA)

All rights reserved.

No part of this publication may be reproduced or transmitted in any form or by any means, electronic or mechanical, including photocopy, recording, or any information storage and retrieval system, without permission in writing from the Publisher.

The appearance of the code at the bottom of the first page of a chapter in this book indicates the Publisher's consent that copies of the chapter may be made for personal or internal use of specific clients. This consent is given on the condition, however, that the copier pay the stated per copy fee through the Copyright Clearance Center, Inc. (222 Rosewood Drive, Danvers, Massachusetts 01923), for copying beyond that permitted by Sections 107 or 108 of the U.S. Copyright Law. This consent does not extend to other kinds of copying, such as copying for general distribution, for advertising or promotional purposes, for creating new collective works, or for resale. Copy fees for pre-2003 chapters are as shown on the title pages. If no fee code appears on the title page, the copy fee is the same as for current chapters. $35.00

Explicit permission from Academic Press is not required to reproduce a maximum of two figures or tables from an Academic Press chapter in another scientific or research publication provided that the material has not been credited to another source and that full credit to the Academic Press chapter is given.

ACADEMIC PRESS
An Elsevier Science Imprint
525 B Street, Suite 1900, San Diego, California 92101-4495, USA
http://www.academicpress.com

International Standard Book Number: 0-12-044981-1

Printed in the United States of America

02 03 04 05 06 07 08 MM 9 8 7 6 5 4 3 2 1

CONTENTS

PREFACE ix

1 Relating Chromatography to Separations

    I. Defining Separation 2
    II. Evolution of Chromatography 3
    III. Separations in Everyday Life 6
    IV. Basis of Separations 7
    V. Modes of Chromatography 13
    VI. Unified Separation Science 14
    VII. Selectivity and Detectability 15
    References 16
    Questions for Review 16

2 Simple Separation Methods

    I. Evaporation 18
    II. Precipitation 18
    III. Crystallization 19
    IV. Filtration 19
    V. Membrane Separations 21
    VI. Distillation 24
    VII. Extraction 26
    References 35
    Questions for Review 35

3 Equilibrium Processes in Separations

   I. Molecular Interactions   38
  II. Separation Thermodynamics   38
      References   48
      Questions for Review   48

4 The Molecular Basis of Separation

   I. Molecular Interactions   49
  II. Solubility Parameter Theory   60
 III. Group Interactions   65
      References   67
      Questions for Review   67

5 Mass Transport and Separation

   I. Types of Diffusion   70
  II. Modeling Diffusion   71
 III. Chromatographic Implications   72
  IV. Diffusion Rates in Various Media   74
   V. Mass Transfer   77
      References   79
      Questions for Review   79

6 Chromatographic Methods

   I. General Classification of Chromatographic Methods   84
  II. Classification Based on Retention Mode   87
 III. Classification Based on Sample Introduction   90
  IV. Separation Characteristics of Chromatographic Methods   93
   V. Basic Chromatographic Theory   95
      References   99
      Questions for Review   100

7 Paper Chromatography

    I. Chromatography Paper   102
   II. Sample Preparation   103
  III. Sample Cleanup   104
   IV. Derivatization   105
    V. Mobile Phases and Stationary Phases   106
   VI. Development of Chromatograms   110
  VII. Detection   111
 VIII. Quantitation   112
       References   112
       Questions for Review   112

8 Thin-Layer Chromatography

I. Stationary Phases for TLC   114
II. TLC of Enantiomeric Compounds   117
III. Sample Application   119
IV. Mobile Phases   123
V. Development of Chromatograms   124
VI. Detection and Quantitation   125
VII. Applications   126
   References   131
   Questions for Review   131

9 Gas Chromatography

I. Equipment   134
II. Separation Process   135
III. Columns   136
IV. Gas Chromatographic Detectors   142
V. Recording and Analysis   145
VI. Resolution   146
VII. Selection of a Stationary Phase   147
   References   152
   Questions for Review   152

10 High-Pressure Liquid Chromatography

I. Evolution of HPLC   154
II. Advantages over Gas Chromatography (GC)   155
III. Separation Process   157
IV. Retention Parameters in HPLC   160
V. Resolution and Retention Time   163
VI. Equipment   167
VII. Separation Mechanism in HPLC   168
VIII. Stationary Phase Effects   179
IX. A Case Study of Retention Mechanism Investigations   180
X. Molecular Probes/Retention Index   182
XI. Mobile Phase Selection and Optimization   185
   References   205
   Questions for Review   208

11 Evolving Methods and Method Selection

I. Evolving Methodologies   211
II. Selection of a Separation Method   220
III. Chiral Separations   223
IV. Comparison of GC, SFC, HPLC, and CEC for a Selective Separation   239
   References   240
   Questions for Review   241

Index   243

PREFACE

For more than 25 years, I led a group in Analytical R&D in a major pharmaceutical company (Novartis), and also taught a course on chemical separations at Pace University. I had the pleasure of succeeding Lloyd Snyder, who had taught this course before me. We used Karger, Snyder, and Horvath's *An Introduction to Separation Science* (Wiley) as a text for this course. I also taught graduate courses on separations and advanced biochemical analysis at Rutgers University and a course on modern methods of chemical analysis at Polytechnic University. I found that I had to use at least a dozen books to convey the subject of chromatography and separation science to the students. All of those books and more are adequately referenced in this text to enable the reader to seek more details. The books that were found especially useful, besides the one mentioned above, were Pecosk, Shields, Cairns, and McWilliam's *Modern Chemical Analysis* (Wiley), Stahl's, *Thin-Layer Chromatography* (Springer-Verlag), Snyder and Kirkland's *Introduction to Modern Liquid Chromatography* (Wiley), Poole and Poole's *Chromatography Today* (Elsevier), McNair and Miller's *Basic Gas Chromatography* (Wiley), Miller's *Separation Methods in Chemical Analysis* (Wiley), Giddings's *Dynamics of Chromatography* (Marcel Dekker) and *Unified Separation Science* (Wiley), and books by me, entitled *Selectivity and Detectability Optimizations in HPLC* (Wiley), *Trace and Ultratrace Analysis by HPLC* (Wiley), and *Chiral Separations by Chromatography* (Oxford). I am indebted to the valuable contributions made by these scientists.

After several years of dealing with this number of books, it occurred to me to plan a text that would cover this subject of chromatography and separation science in a concise manner and avoid burdening the reader with a lot of details and mathematical equations. This book is designed for students and separation scientists who are not averse to learning a few equations dealing with physicochemical processes that influence chromatography and separations. The main objectives of this book are as follows:

- Provide basic information on chromatography and separation science.
- Describe the relationship between these important fields.
- Cover how simple extraction or partition processes provide the basis for development of chromatography and separation science.
- Describe the role of chromatography and separation science in various fields.
- Discuss the role of chromatography and separation science in development of new methodology.
- Cover new evolving methods and show how to select an optimum method.

Separation can be defined as an operation in which a mixture is divided into at least two fractions having different composition, molecular mass, or stereochemical structure. An actual separation is usually achieved by physical process, although chemical reactions may occasionally be involved in this process. It is important to learn the fundamental physical and chemical phenomena involved in the achievement of separations, as well as those with the development and application of various separation processes. This continually evolving discipline is called *separation science*.

Chromatography derives its name from *chroma* and *graphy,* meaning color writing. It is essentially a physical method of separation in which components to be separated are distributed between two phases. Chromatography is a powerful technique concerned with separations of a large variety of complex compounds. The contributions of chromatography to various scientific disciplines and the benefits that chromatography provide to mankind are unparalelled. For example, the progress made in the biological sciences such as biotechnology, clinical pharmacology/therapeutics, and toxicology provides an excellent testimony to the contributions of chromatography. Examples from these and other fields are included throughout this textbook.

This book will provide the necessary information to improve the comprehension of separation science. It will show how chromatography relates to separation science and describe how simple separation methods have led to highly useful chromatographic techniques that provide high resolution (Chapters 1 and 2). To provide a better understanding of transport phenomena and thermodynamics, an elementary discussion is included in Chapters 3 to 5.

The discussion on chromatography should help improve the understanding of the principles involved in various modes of chromatography such as paper chromatography, thin-layer chromatography, gas chromatography, and high-pressure liquid chromatography (Chapters 6 to 10).

This book relates simple separation techniques—such as extraction—to chromatography. And it lets us see how chromatography in turn relates to capillary electrophoresis or field flow fractionation (Chapter 11).

I believe that both theory and many interesting applications of various chromatographic and separation techniques included in this book will be found useful by the readers.

November 6, 2002   Satinder Ahuja

1 RELATING CHROMATOGRAPHY TO SEPARATIONS

I. DEFINING SEPARATION
II. EVOLUTION OF CHROMATOGRAPHY
   A. Definition of Chromatography
   B. Similarity of Chromatography to Separation Methods
III. SEPARATIONS IN EVERYDAY LIFE
IV. BASIS OF SEPARATIONS
   A. Physicochemical Phenomena
   B. Utilizing a Desirable Physical Property
V. MODES OF CHROMATOGRAPHY
   A. Adsorption Versus Absorption
   B. Partition or Distribution
   C. Exclusion
   D. Ion Exchange
VI. UNIFIED SEPARATION SCIENCE
VII. SELECTIVITY AND DETECTABILITY
   A. Selectivity
   B. Detectability
   REFERENCES
   QUESTIONS FOR REVIEW

A chromatographic method can be considered simply a physical method of separation in which components to be separated are distributed between two phases. However, it should be recognized that chromatographic methods have evolved into complex and elegant methods that utilize a variety of physicochemical approaches to provide the desired separation of complex molecules (see Chapters 6–10). Chromatography can be related directly to some of the simple nonchromatographic separation methods by the same common basic physicochemical principles employed in these methodologies to achieve the desired results. We will see from the discussion in Sections I and II that separation is a broad term that can include nonchromatographic methods as well as all of the chromatographic methods, and it should be recognized at the very outset that this does not imply that all separation methods are chromatographic.

Chromatography provides a variety of powerful methods that can help separate a large number of different compounds. As a matter of fact, it can be said without much hesitation that a chromatographic method is likely to be the method of choice for complex samples when a selection has to be made among various separation methods. The contributions of chromatography to various scientific disciplines are numerous. The investigations conducted with the chromatographic methods have resulted in a great variety of benefits to human beings. A large number of examples can be found in the scientific literature of chemical and biological sciences such as biochemistry, biotechnology, clinical pharmacology, therapeutics, and toxicology. Most importantly, the progress made in discovery of new drugs and new approaches to treat diseases provides an excellent testimony to the contributions of this methodology.[1,2] This progress can be related to unparalleled separation power offered by chromatography. Examples from these and other fields are included throughout this book.

A discipline that encompasses the art and science of separation has been evolving for some time now. It deals with the fundamental physical and chemical phenomena involved in the achievement of separations, as well as with the development, application, and reproducibility of various separation processes. This discipline can be appropriately called separation science and is quite heavily dependent upon physicochemical principles involved in separations, that is, chemistry of separations. It requires that we improve our understanding of the common principles underlying various separation processes (see Chapters 3–5). This means we should develop better understanding of transport phenomena and thermodynamics up to at least the elementary level.

This book has been planned primarily to help readers gain a better understanding of the common principles underlying various separation processes and thus help to make better use of these powerful methods. In this text, fairly detailed discussions of both the fundamental aspects and the practical applications of separations, including chromatography, are presented, with major emphasis on laboratory and analytical-scale separations. Small-scale nonchromatographic and chromatographic separations that are commonly used by many chemists and employed very frequently in various other areas of science and technology have also been discussed to demonstrate the diversity and the unity of separation science. A few of these separation processes that have been found useful in large-scale industrial applications have been included. It should be noted that chromatography is covered at length in this book because it is now a very important technique for solving a large number of different kinds of problems in separation science.

I. DEFINING SEPARATION

Since separations are important in almost all our endeavors in science and technology, the word separation is well known, and virtually anyone working in this field can come up with a meaning for this word. However, it is quite difficult to define separation precisely and completely.[3–6] Separation has been simply defined as an operation in which a mixture is divided into at least two components having different compositions. However, this definition does not cover chiral separations where molecules have the same composition and chemical structure and differ only in their stereochemistry. So a broad definition that would overcome this shortcoming is as follows:

Separation is an operation in which a mixture is divided into at least two components with different compositions, or two molecules with the same composition but different stereochemical structure.

The separation process usually entails physical means. A number of simple methods that are utilized for various separation processes have been discussed in Chapter 2. Flicking a fly from a glass of water is a good example of separation. In practical terms, this process can be described as a separation of a solid matter (viz., the fly, which can be described as a particle of a certain size) from the liquid by the mechanical means. The same separation could have been achieved by scooping the fly with some device such as a glass rod, or filtering the liquid through a filter paper or a piece of cloth. These separation processes do not eliminate any bacterial contamination. To achieve sterility by eliminating microbial contamination, one would have to use specialized filters or other suitable techniques.

Most separations are not this simple in real life because the particles of interest may get smaller—to a point that they are invisible to the naked eye—and the materials to be separated may be solubilized in the solvent from which one is trying to achieve separation.

II. EVOLUTION OF CHROMATOGRAPHY

Chromatography derives its name from *chroma* and *graphy*, meaning color writing. As a matter of fact, chromatography owes its origin to the efforts of the Russian botanist M. S. Tswett at around the turn of the twentieth century.[7] He used a column of powdered calcium carbonate to separate green leaf pigments into a series of colored bands by allowing a solvent to percolate through the column bed. This technique can be simply described as liquid–solid adsorption chromatography on a column. Historically, the origin of chromatography can be traced back to 1905 when W. Ramsey separated a mixture of gases and vapors from solid sorbent, such as active charcoal.[8] This technique entailed adsorption and can be related to adsorption chromatography; a forerunner of gas–solid chromatography (see Chapter 9).

The logic of naming the technique used by Tswett as chromatography does not hold true today because most of the compounds separated by this technique are not colored and detection is not based on observation of color (see Chapter 6 for more details). However, the use of this name is now fully established and is not likely to change. The technique remained virtually dormant until the early 1940s, when the well-known paper of Martin and Synge was published.[9] They reported the discovery of liquid–liquid partition chromatography, both on columns and on paper. The latter technique is more commonly called paper, or planar, chromatography (see Chapter 7). Martin and Synge also provided a theoretical framework for the basic chromatographic process, and they received the Nobel Prize in chemistry in 1952 for their work. As a result, their discoveries had a major impact in this area, especially in biochemistry.

The next major step that led to progress in this field was the development of gas–liquid chromatography by James and Martin.[10] This technique differed from liquid chromatography in that compressibility of the gaseous mobile phase was utilized in a specific way (see Chapter 9). The introduction of gas–liquid chromatography had an unprecedented impact on analytical chemistry of organic compounds.

Porath and Flodin[11] introduced gel filtration, a method for group separation of compounds. This led to gel filtration or steric exclusion chromatography that allows easy separation of macromolecules.

The success of modern chromatography is greatly due to the excellent extensive treatment of chromatographic theory by Giddings in 1965 in his book entitled *Dynamics of Chromatography*.[12] A number of well-known scientists whose contributions are too numerous to be recounted here are mentioned in this book in appropriate places (see Chapters 6–11), and their work has led to the development of modern liquid chromatography, which is often called high-pressure or high-performance liquid chromatography. This technique is also called HPLC, or simply LC.

Other modern chromatographic techniques such as supercritical fluid chromatography and capillary electrochromatography (an interesting by-product of electrophoresis and chromatography) are discussed later in this book.

A. Definition of Chromatography

Based on the preceding discussion, chromatography can be defined as follows:

Chromatography is essentially a physical method of separation in which components to be separated are distributed between two phases, one of which does not move (appropriately called the stationary phase) and the other that moves through it in a definite direction (commonly described as the mobile phase).

B. Similarity of Chromatography to Separation Methods

Various modes of chromatography have been discussed in this book. For example, the column chromatography first discovered by Tswett, can be related to liquid–solid adsorption. Of the various other modes of chromatography described in Section IV, it is easy to see a relationship between the separation method called liquid–liquid extraction and the well-known chromatographic technique called liquid–liquid partition chromatography or, more simply, liquid chromatography. The separation process in liquid chromatography is similar to liquid–liquid extraction, where two phases are immiscible liquids.

The extraction process can be explained as follows: When the two liquids are shaken together to achieve an extraction of a component of interest in a separatory funnel (Fig. 1), the sample proportionates itself into the two phases based on its distribution coefficient:

$$\text{Distribution coefficient } (K) = \frac{\text{concentration of sample in phase 1}}{\text{concentration of sample in phase 2}}$$

The two phases in extraction, in technical jargon in industry, are called the raffinate and the extractant. The stationary phase in chromatography is similar in function to the raffinate in extraction and the mobile phase is equivalent to the extractant. It may be instructive to note at this point that in dialysis, another separation technique, one or more solutes are transferred from one fluid to another through a membrane under a concentration driving force (see Chapter 11). The membrane separates the two phases, and the phase that is retained is called retentate, and the one that diffuses through the membrane is appropriately designated as diffusate.

Now getting back to extraction: If one of the phases is removed and a fresh amount of the same phase is added, one can extract more of the sample into the newly added portion from the phase that was not removed, that is, the stationary phase. This is similar to using a mobile phase in chromatography, where the mobile phase is always fresh. The utilization of a fresh mobile phase that is moving continuously through and around the selected stationary phase provides the basis for development of various chromatographic techniques.

In various other modes of chromatography discussed in this book, the terms mobile phase and stationary phase are used to describe the two phases. However, it should be noted that neither of them necessarily has to be liquid. For example, in gas–solid chromatography, the mobile phase is gas and the stationary phase is solid.

It should be apparent by now that even though chromatographic methods can be fitted under the broad category of separation, not all separation methods are chromatographic. In chromatography the two phases are utilized in a number of innovative ways to provide enormous advantages that help resolve a large variety of compounds (see Chapters 6–10).

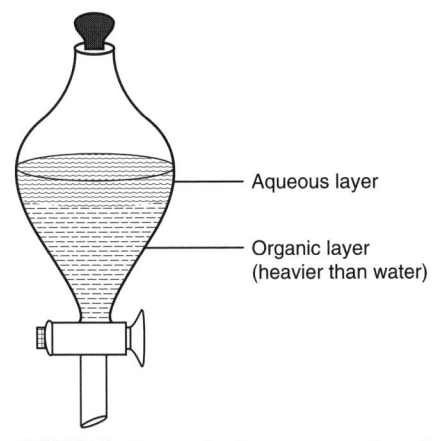

FIGURE I Extraction in separatory funnel.

III. SEPARATIONS IN EVERYDAY LIFE

We utilize separation methods virtually all the time. For example, we use air filters to allow us to breathe clean air indoors. Depending on the type of filter used, we can remove relatively large dust particles or microscopic pollen from plants that may produce many types of allergies. Similarly, our municipal drinking water goes through various purification steps to separate out the impurities. It is important for us to understand the chemistry of separations to better deal with our purification problems in various human endeavors.

Chemistry is a discipline that is based on a variety of separation processes. It plays a major role in our lives. As a matter of fact, it is interesting to note that *scheikunde*, the Dutch word for chemistry, means the art of separation. From time immemorial, humanity has made efforts to better understand its environment (air, water, food, and other elements that affect us directly or indirectly), to protect our health, and to make changes that would benefit people at large. These efforts have been closely linked to the separation of chemical substances.

It may be interesting to review a number of examples of historical interest that relate to some process of chromatographic or nonchromatographic separation. Moses' conversion of bitter water into sweet drinkable water by the immersion of tree branches may have been the first recorded application of ion-exchange separations.[3] Other well-known examples are listed below.

1. Distillation of alcohol for various purposes is a simple separation process.
2. Isolation of dyes to add color to various materials to add sparkle to our life depends on various separation processes that include chromatography.
3. Extraction of natural products to provide useful drugs relies very heavily on chromatography.
4. Isolation of metals in metallurgy is based on various techniques of separation.

Industrial and scientific revolutions have brought about the need for development of a variety of separation methods, and this need is being met admirably by separation science. As a matter of fact, most advances in chemistry, chemical technology, and life sciences have been related to advances in separations and chromatography. A few examples follow:

1. The automobile industry and transportation owe their success to the effective use of separation processes used in the petroleum industry; they have truly revolutionized our lives today.
2. The atomic age owes its progress to the resolution of the difficult problem of separation of ^{235}U from ^{238}U.
3. The studies on separation processes in the laboratory are of great significance in understanding the phenomena taking place in vivo, such as selective permeation of cell membranes. These studies can help us determine how diseases are transmitted and how we can treat them better with suitable medicines.

4. Our understanding of a variety of biochemical processes occurring in the human body became possible after development of separation methods such as ultracentrifugation, chromatography, and electrophoresis.
5. The studies on the human genome and proteomics owe their success to separation science. This knowledge will lead to better treatment of diseases.
6. The use of the artificial kidney is a notable example of the use of separation processes in modern medicine.
7. To assure the preparation of contaminant-free materials needed for various industrial processes, it is necessary to use highly sensitive chromatographic methods for trace and ultratrace analysis (discussed later in this section).
8. Forensic analysis to monitor arson or explosives used by terrorists also demands the use of trace and ultratrace chromatographic analysis.
9. The future solutions of environmental problems relating to air, water, and food will depend to a great extent on finding economical ways to solving the corresponding separation problems.

It goes without saying that the removal of contaminants from the soil, water, air, food, or other items consumed by human beings will continue to require application of separation processes on a large scale. At the same time, it will be necessary to monitor and control the introduction of new pollutants by rigorous development of trace and ultratrace monitoring methods based on separations.[13]

Trace analysis generally entails analysis at parts per million level (ppm); for example, analysis at 1 ppm level equates to 0.0001% and requires determination of 1 microgram of a component of interest in 1 gram of a sample. Ultratrace analysis is generally performed at well below the parts per million level. As a matter of fact, it is not uncommon to carry out these analyses at parts per billion (one nanogram/gram) or parts per trillion (one picogram/gram) or even lower levels.

IV. BASIS OF SEPARATIONS

A number of methods can be used for separation of particles. These methods are often referred to as mechanical separation processes. A well-known example is screening, or elutriation, where particles are separated according to their size and shape. The methods that can be used for separation of particles are listed here alphabetically.

- Electrostatic precipitation
- Elutriation
- Field flow fractionation
- Filtration
- Flotation
- Particle electrophoresis
- Precipitation

- Screening
- Sedimentation
- Ultracentrifugation

As mentioned before, the simplest form of separation entails removal of particles that are visible. However, the methods become more complex as the particle size decreases, since there is no sharp break between very large molecules, colloid particles, and macroscopic particles.

The separation methods are often described by and named after underlying forces or phenomena that give rise to fractionation; for example, precipitation, crystallization, sedimentation, centrifugation, extraction, adsorption, ion exchange, diffusion, and thermal diffusion. At times, the name applies to a distinct form of operation, such as filtration, distillation, dialysis, zone melting, thermogravitational separations, elutriation, field flow fractionation, electrostatic precipitation, chromatography, and so on. Chromatography, for example, can employ any of a number of forces such as adsorption, ion exchange, extraction, and exclusion (see Section V).

A large number of separation methods are available that utilize select characteristics as means of separation. Each of these methods can be further divided into different techniques with unique characteristics. The need for a great variety of separation procedures can be attributed to the following factors:

- Different separation goals
- Diversity of the mixtures to be separated
- Variety of physicochemical phenomena involved in separation

A common objective of various separation methods is adequate segregation of constituents of interest with maximum speed and minimum effort, and with as large a capacity as possible. Separation plays an important role in the analysis of mixtures, as exemplified in the following examples.

- Removal of interfering constituents before the determination of one or more known compounds
- Isolation of unknown compounds for further characterization
- Analysis of complex unknown mixtures by separation of the entire sample into individual constituents

Mixtures to be separated can vary greatly with respect to molecular weight, volatility, or other properties. They may range from mixtures of atomic species through organic molecules and macromolecules to molecular aggregates (particles). In some cases the properties of the individual components are so different that very simple separation techniques can be used (e.g., salt can be recovered from seawater by simply evaporating water, or it can also be collected by condensation in a distillation unit). But in other cases the species to be separated may be scarcely distinguishable (e.g., pairs of optical isomers). The complexity of the sample mixture with respect to the number of constituents or their physical and chemical properties affects the difficulty of separation and the type of technique that can be utilized. Most important, the choice of a separation technique depends on the amount of mixture to be separated, which can range from a few molecules that would require microtechniques suitable for ultratrace analysis to tons of material in an industrial-scale process that would require preparative-scale separations.

A. Physicochemical Phenomena

A large number of physicochemical phenomena have been used to achieve various separations. To simplify, we can divide these phenomena into two major categories, equilibrium processes and rate processes. These processes are described briefly here; more in-depth discussion is provided later in the book.

1. Equilibrium Processes

Equilibrium processes are based on differences in the properties of individual components to be separated. Equilibrium separation processes are often based on phase equilibria. The most common equilibrium phenomena involve the distribution of substances between two phases:

- Gas–liquid
- Gas–solid
- Liquid–liquid
- Liquid–solid

It should be noted that both chromatographic and nonchromatographic methods can utilize equilibrium between two phases as the basis of separations.

2. Gas–Liquid Processes

Some examples of separation methods where the two phases are gas and liquid are shown here.

1. Nonchromatographic Methods
   (a) Distillation
   (b) Foam fractionation
2. Chromatographic Methods
   (a) Gas–liquid chromatography

3. Gas–Solid Processes

A few examples of separation methods where the two phases are gas and solid are shown below.

1. Nonchromatographic Methods
   (a) Sublimation
   (b) Adsorption
2. Chromatographic Methods
   (a) Gas–solid chromatography
   (b) Exclusion chromatography

Sublimation, which is a form of distillation where a solid is directly vaporized, is dependent on the equilibrium between the gas and solid phases. Adsorption of gases by solids has provided an excellent basis for development of gas–solid chromatography. Exclusion chromatography, where molecules are separated based on their size or shape, can be performed with either a gas or a liquid phase.

4. Liquid–Liquid Processes

The most notable example of a nonchromatographic method based on equilibrium between liquid–liquid phases is extraction. We know from earlier discussion that extraction is dependent on favorable partition of the desired component into one of the two liquid phases, which are immiscible with each other.

1. Nonchromatographic Methods
   (a) Extraction
2. Chromatographic Methods
   (a) Liquid–liquid column chromatography
   (b) Exclusion chromatography
   (c) High-pressure liquid chromatography (HPLC)

5. Liquid–Solid Processes

Precipitation is a well-known nonchromatographic method shown here that is based on equilibrium between the solid and liquid phases. Other methods in this group are fractional crystallization and clathration (where one type of molecule or a framework of one type of molecule encloses a different molecule).

1. Nonchromatographic Methods
   (a) Precipitation
   (b) Fractional crystallization
   (c) Clathration
2. Chromatographic Methods
   (a) Adsorption chromatography
   (b) Ion-exchange chromatography
   (c) Exclusion chromatography

Of the various methods described above, the chromatographic methods (GC or HPLC) are most commonly used to solve some complex problems.

At times, an equilibrium distribution in a single phase, resulting from the action of a force field, can also be used for separation, for example, equilibrium ultracentrifugation.

6. Rate Processes

Rate processes are based on differences in the kinetic properties of the components of mixtures. These can encompass the following:

- Diffusion rates through permeable barriers such as membranes
- Migration velocities in various fields such as gravitational, electrical, and thermal
- Miscellaneous physical or chemical processes

7. Diffusion Through a Permeable Barrier

The most common permeable barrier separation method is filtration through a filter paper. A number of other methods that have evolved over a period of time and use permeable membranes are listed alphabetically here.

- Dialysis
- Electrodialysis
- Electroosmosis
- Gaseous diffusion
- Membrane filtration
- Reverse osmosis
- Ultrafiltration

8. Field Separations

Electophoresis and its more recent variant, capillary electrophoresis (see Chapter 11), are well-known examples of field separations. In these cases an electrical field is applied to achieve the desired separation. A number of field separation methods are listed here alphabetically.

- Capillary electrophoresis
- Electrodeposition
- Electrophoresis
- Mass spectrometry
- Thermal diffusion
- Ultracentrifugation

An impressive example of field separations is mass spectrometry. It involves the production of gas-phase ions from a sample and the resulting separation according to their mass to charge (m/e) ratio. Interestingly, mass spectrometry is very rarely looked upon as a separation method even though it separates molecules based on their molecular mass. It can provide valuable information as follows:

- Elemental composition of the sample
- Structure of organic and inorganic molecules including molecules of biological interest
- Qualitative and quantitative composition of complex mixtures
- Structure and composition of solid surfaces
- Isotopic ratio of atoms in the sample

It should be recognized that mass spectrometers are used primarily by separation scientists as detectors. Further discussion on mass spectrometry as a detector for gas chromatography, high-pressure liquid chromatography, and capillary electrophoresis has been included with the techniques listed.

9. Miscellaneous Separation Methods

A few miscellaneous separation methods are listed here alphabetically.

- Destructive distillation
- Enzymatic degradation
- Molecular distillation

Most of the separation methods utilize some physical property (or properties) to achieve the desired separation (see Section IV. B). However, some of the separation processes can involve reactions based on chemical properties; for example, destructive distillation and enzymatic degradation.

B. Utilizing a Desirable Physical Property

Another basis of classification of separations is a certain desirable property such as:

- Volatility
- Solubility
- Partition coefficient
- Ion exchange
- Surface activity
- Molecular geometry
- Electromigration

1. Volatility, Solubility, or Partition Coefficient

Distillation requires that the components of a mixture be volatilized to yield the desired separation. It is generally necessary to use heat to achieve this separation. Similarly, it is necessary to volatilize a sample to achieve separation in gas chromatography; however, the partition process is primarily responsible for the separation.

Zone refining requires the use of heat to achieve useful separations. It is important to recognize that in this case solubility is the primary useful property that provides the desired separation.

The methods based on partition require a second phase. Some examples are as follows:

- Extraction
- Liquid–liquid partition chromatography
- Gas–liquid chromatography

The simplest example of partition is liquid–liquid extraction. The methods based on partition have evolved into useful chromatographic methods (see classification of chromatographic methods in Section V). Partition chromatography can be simply performed on gravity-fed columns, where one of the partitioning phases (stationary phase) remains on the column and the mobile phase flows down. Notable examples of partition chromatography are gas–liquid chromatography (GLC) and high-pressure liquid chromatography (HPLC), where high pressure is used to pump a gaseous or liquid mobile phase, respectively. It should be noted that in GLC it is necessary to volatilize the sample before one can take advantage of partition. When the surface activity of the adsorbent on a column plays a major role, the chromatographic technique is called adsorption chromatography.

2. Ion Exchange

Ion-exchange equilibria can lead to separation of cations or anions. Chromatographic methods have been found very useful for this purpose. In ion-exchange chromatography, exchange equilibrium of the ions plays a major role.

3. Surface Activity

The methods that utilize surface activity are adsorption methods. They can be nonchromatographic or chromatographic. Some of the chromatographic methods are as follows:

- Liquid–solid adsorption chromatography
- Gas–solid adsorption chromatography

4. Molecular Geometry

The following methods based on molecular geometry require a permeable barrier to provide successful separations: molecular sieves, gel filtration, gas diffusion, inclusion complexes, ultrafiltration, dialysis, and electrodialysis.

5. Electromigration

The success of electrophoresis and the evolution of capillary electrophoresis and electrochromatographic methods relate to electromigration properties of the sample.

V. MODES OF CHROMATOGRAPHY

Based on the previous discussion, chromatographic methods can be broadly and simply classified as follows, based on two phases involved in the separation:

1. Adsorption chromatography
   (a) Gas–solid adsorption
   (b) Liquid–solid adsorption
2. Partition chromatography
   (a) Gas–liquid partition
   (b) Liquid–liquid partition
3. Exclusion chromatography
   (a) Gas–solid exclusion
   (b) Liquid–solid exclusion
4. Ion-exchange chromatography
   (a) Ion exchange on resinous materials, cellulosic, or inorganic materials

A. Adsorption Versus Absorption

At this time, it important to distinguish between *adsorption* and *absorption*. Adsorption relates to distribution processes occurring at the interface, distinguishing it from absorption, which describes distribution processes occurring in bulk phases.

B. Partition or Distribution

These are similar terms and are sometimes used interchangeably. They relate to the general process by which a system comes to equilibrium.

C. Exclusion

This relates to separations based on size or shape, where certain molecules are excluded from the stationary phase, resulting in separation.

D. Ion Exchange

This process relates to exchange of a given ion, cation, or anion during the separation process.

VI. UNIFIED SEPARATION SCIENCE

Giddings[14] suggested that a large number of techniques that are commonly used for chemical separations can be unified by formulating the concepts that unite them. Since all chemical separations require the movement of components from a region commonly shared to an individual region, the processes of differential displacement must underlie the entire methodology. Therefore, a separation process has to fundamentally entail a selective movement through space of one component with respect to another. From this short discussion, we can conclude that the science of separations is rooted primarily in transport phenomena. It is important to understand the underlying mass transport phenomena of all separation processes. This will allow us to describe and explore the limitations and performance of alternative separation techniques and the factors enhancing and limiting separation power.

Separation procedures will continue to change and evolve with time; a better understanding of the underlying theory and mechanism of separations will help us make better use of these methods and allow us to think beyond using them merely as recipes.

Giddings[14] maintained that all separation systems could be fitted into nine categories as shown in Table 1, based on whether

- Chemical potential profile, μ^*, (representing the sum of external field effects and internal molecular interactions) is continuous (c).
- Chemical potential profile, μ^*, is discontinuous (d).
- Chemical potential profile, μ^*, is a combination of both (cd).
- Flow is static.
- Flow and μ^* gradient are parallel.
- Flow and μ^* gradient are perpendicular.

I RELATING CHROMATOGRAPHY TO SEPARATIONS

TABLE I Classification of Separation Methods

Flow	Continuous profile	Discontinuous profile	Combination profile
Static	Electrophoresis	Evaporation	Electrodeposition
	Isoelectric focusing	Crystallization	Electrostatic precipitation
	Isotachophoresis	Extraction	Electrolytic refining
	Rate zonal sedimentation	Distillation	Electrodialysis
		Adsorption	Equilibrium sedimentation
		Sublimation	
		Ion exchange	
		Dialysis	
Flow and μ^* parallel	Elutriation	Filtration	Electrofiltration
	Countercurrent electrophoresis	Ultrafiltration	
		Reverse osmosis	
		Pressure dialysis	
		Zone melting	
Flow and μ^* perpendicular	Hyperlayer field-flow fractionation	Foam fractionation	Electrodecantation
		Countercurrent distribution	Thermogravitational separation
		Fractional distillation	Field-flow fractionation

Source: Adapted from Reference 14.

The important point to remember is that each of these methods offers different selectivity.

VII. SELECTIVITY AND DETECTABILITY

The desirability of a given separation method relates to the selectivity of separations offered by it.[6,13]

A. Selectivity

Selectivity of a method relates to its ability to separate components with slight variations in molecular weight or structure Selectivity also relates to the fundamental physiochemical phenomena underlying a separation. It refers to the intrinsic capability of a given separation method to distinguish between two components. The difference in molecular weight may be as slight as the replacement of hydrogen atoms with deuterium atoms, that is, compounds that differ in isotopic composition. For example, deuterated polyaromatic hydrocarbons can be separated from their protiated analogs. Other examples of high selectivity relate to separation of compounds that differ only in the position of a double bond. For compounds with the same molecular weight, the structural difference may involve no more than one of the stereochemical forms, viz., geometric or optical isomers.[15,16] And differences in structure may be no more than that of a molecule and its mirror image, as exemplified by optical isomers.

B. Detectability

Detectability relates to a minimum quantity of material that can be detected. This quantity can be calculated based on certain assumptions, such as detection at two or three times the noise level seen at the baseline, or it can be determined practically. Unquestionably, detectability relates to separation efficiency, minimum band broadening, and sensitivity of detectors. A good knowledge of separation processes can help us improve detectability by minimizing band broadening, a common problem in separation processes, even when the same detector is being used.

To optimize selectivity and detectability, it is necessary to understand the physicochemical basis of retention. An attempt has been made in this book to clarify the basis of these phenomena and to provide a number of approaches to optimize selectivity and detectability of various methods.

REFERENCES

1. Ahuja, S. *Chromatography and Separation Chemistry*, ACS Symposium Series #297, American Chemical Society, Washington, DC, 1986.
2. Ahuja, S. *Chromatography of Pharmaceuticals: Natural, Synthetic, and Recombinant Products*, ACS Symposium Series #512, Washington, DC, 1992.
3. Karger, B., Snyder, L., and Horvath, C. *An Introduction to Separation Science*, Wiley, New York, 1973.
4. Snyder, L., Glajch, J., and Kirkland, J. *Practical HPLC Method Development*, Wiley, New York, 1988.
5. Poole, C. and Poole, S. *Chromatography Today*, Elsevier, New York, 1991.
6. Ahuja, S. *Selectivity and Detectability Optimization in HPLC*, Wiley, New York, 1989.
7. Tswett, M. S. *Ber. Deut. Botan. Ges.* 24:316 and 384, 1906.
8. Ramsey, W. *Proc. Roy. Soc.* A76:111, 1905.
9. Martin, A. J. P. and Synge, B. L. M. *Biochem. J.* 35:1358, 1941.
10. James, A. T. and Martin, A. J. P. *Biochem. J.* 50:679, 1952.
11. Porath, J. and Flodin, P. *Nature* 183:1657, 1959.
12. Giddings, J. C. *Dynamics of Chromatography*, Part I, Marcel Dekker, New York, 1965.
13. Ahuja, S. *Ultratrace Analysis of Pharmaceuticals and Other Compounds of Interest*, Wiley, New York, 1986.
14. Giddings, J. C. *Unified Separation Science*, Wiley, New York, 1991.
15. Ahuja, S. *Chiral Separations: Applications and Technology*, American Chemical Society, Washington, DC, 1997.
16. Ahuja, S. *Chiral Separations by Chromatography*, Oxford, New York, 2000.

QUESTIONS FOR REVIEW

1. Define separation and chromatography.
2. How are chromatography and separations related to each other?
3. What differentiates chromatographic methods from related separation methods?
4. Define selectivity and detectability.

2 SIMPLE SEPARATION METHODS

    I. EVAPORATION
    II. PRECIPITATION
        A. Solvent Precipitation
        B. Precipitation via Chemical Reaction
        C. Precipitation by Adjustment of pH
    III. CRYSTALLIZATION
    IV. FILTRATION
    V. MEMBRANE SEPARATIONS
        A. Structure and Properties of Various Membranes
        B. Membrane-Based Separation Methods
    VI. DISTILLATION
        A. Phases in Distillation
        B. Modes of Distillation
        C. Distillation as the Precursor to Gas Chromatographic Methods
    VII. EXTRACTION
        A. Distribution Coefficient
        B. Distribution Coefficient of a Species
        C. Distribution Ratio
        D. Successive Extractions
        E. Extraction of Mixtures
        F. Multiple Extractions
        G. Countercurrent Extraction Versus Differential Column System
    REFERENCES
    QUESTIONS FOR REVIEW

Simple methods such as evaporation and filtration have been used by scientists for a very long time to achieve desired separations. These methods as well as a number of other nonchromatographic separation methods that have evolved over a period of time out of these simple methods are listed below. The first method in each group (in bold) may be considered the key method that led to other methods.

- **Evaporation**
- Precipitation
- Crystallization

- **Filtration**
- Ultrafiltration
- Membrane separation
- Dialysis

- Electrodialysis
- Reverse osmosis
- **Distillation** (various modes)
- Sublimation
- **Extraction**
- Countercurrent extraction

A large number of these methods are described briefly in this chapter. It bears repeating that these methods have, in turn, led to many elegant chromatographic and nonchromatographic separation methods that are discussed at length in later chapters of this book (Chapters 6–11). For example, as it was pointed out in Chapter 1, Tswett's discovery of column chromatography is based on liquid–solid adsorption, and liquid–liquid partition chromatographic methods are strikingly similar to the liquid–liquid extraction methods. Both the versatility and limitations of chromatographic methods are included in this text to help readers select the best method for their need.

I. EVAPORATION

Evaporation simply entails vaporizing the solvent by using heat or by utilizing air currents in a manner that the material concentrates to a solid state. For example, common salt (sodium chloride) in a reasonably pure form can be separated from seawater by the simple process of evaporation, using an open container for a period of time or by suitable application of heat.

Evaporation procedures are commonly used in separations to concentrate solutions as a prelude to other separation steps, or they can be the final step for obtaining an isolated material as a residue. The process of evaporation and concentration frequently leads to the formation of crystals that come out of solution. Often, crystalline materials are obtained in a chemistry laboratory or chemical industry by simply concentrating a solution in the selected solvent(s) and allowing the solution to cool. The separation process that leads to formation and isolation of crystals is called crystallization. Evaporation procedures have undeniably led to more useful separation procedures, such as precipitation, crystallization, and distillation, discussed later.

II. PRECIPITATION

Changing the concentration of a solute in a solution so that it exceeds solubility in a given solvent can bring about precipitation of the solute. As mentioned earlier, when this process is carried out in an appropriately selected solvent and in a controlled manner, it can lead to crystallization (see later discussion also). This allows us to obtain pure crystalline materials.

Precipitation can be achieved by a number of different methods. Some of the methods utilized are listed here.

A. Solvent Precipitation

Precipitation can be achieved by adding another miscible solvent such that solubility of the component of interest would be at a low level in the mixed solvent. This procedure is called solvent precipitation.

B. Precipitation via Chemical Reaction

It is possible to obtain a precipitate by chemical means. In this case a change in chemical composition is brought about to obtain a desired material; for example, a precipitate of barium sulfate can be obtained by adding sulfuric acid to an aqueous solution of barium chloride.

$$BaCl_2 + H_2SO_4 \longrightarrow \underset{\text{Precipitate}}{BaSO_4} + 2\,HCl$$

C. Precipitation by Adjustment of pH

Some organic compounds can be precipitated by suitable adjustment of pH. For example, it is possible to precipitate most of the weakly basic organic compounds from aqueous solution by addition of a strongly basic solution. Similarly, the organic acids can be precipitated by acidification of the solution with strong acids.

III. CRYSTALLIZATION

Crystallization is an important procedure used by chemists to obtain pure material in the crystalline state. This involves concentrating a solution containing the component of interest by heating it and then allowing it to stand (i.e., cooling it) until the crystals are obtained from the solution. This process can allow separation of fairly pure material and is utilized for preparation of a large number of chemicals; however, related materials can crystallize simultaneously and thus lead to poor separation. So it becomes necessary to design specialized crystallization procedures from optimal solvents that minimize such contamination.

Diastereoisomers can be separated by forming an appropriate derivative and then crystallizing the desired stereoisomer. However, contamination from other materials crystallizing under the same conditions still remains a major consideration.

IV. FILTRATION

We can reasonably assume that the filtration process has been used from time immemorial to separate solids from liquids. Initially, some porous material (similar to a filter paper or cloth) was most likely used. The primary requirement of such a material is that it should allow liquid, i.e., the solvent,

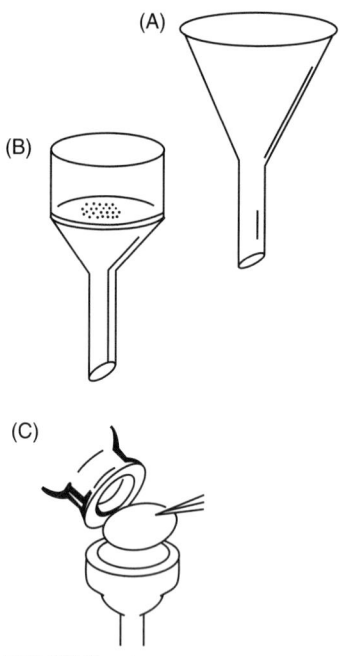

FIGURE 1 Filtration funnels.

to pass through while retaining the solids. This concept has eventually led to a large variety of filter papers and membranes of various pore sizes.

Separation of particles that are visible to the naked eye with a filter paper is called filtration; however, filtration of submicron particle size is also possible (see later discussion). Simple filter papers are made from cellulose and exhibit particle retention levels down to 2.5 µm (e.g., Whatman Grade 5 filter paper). These cellulose filter papers are used for general filtration. A wide choice of flow-rate combinations can be used, depending upon application. A simple filtration process entails pouring the liquid over commonly available fluted filter paper (e.g., Whatman Grade 1 is suitable for particles larger than 11 µm) or a filter paper folded into quadrants that has been placed in a conical glass filter funnel (Fig. 1A), then collecting the filtrate at the other end of the funnel.

In this case gravitational force is adequate to achieve the separation of solids from the liquid. Sometimes, extra force is necessary to obtain separation. This is achieved by laying the filter paper flat on a perforated plate and applying vacuum to the collection flask. This type of funnel is called a Buchner funnel (Fig. 1B).

Membrane filters with various pore sizes provide wide choices in filtration. Membrane holders may incorporate either sealed-in sintered glass or removable stainless steel mesh supports (Fig. 1C). Membrane filters made from various materials can offer various pore sizes that allow separations down to 0.02-µm level (see Table 1). The unique properties of borosilicate glass microfiber allow the manufacture of filters with retention levels extended to the sub-micron range (e.g., Whatman Grade GF/F). These filters can withstand temperatures up to 550°C. Furthermore, they can provide fast

TABLE I Flow Rates of Water at 20° C with Membrane Filters*

Pore size (μm)	Cellulose mixed esters	Cellulose nitrate	Cellulose acetate	Regenerated cellulose	Polyamide	PTFE with polyester support
0.02						0.2
0.2	20	20	20	20	10	20
0.45	40	40	40	50	17	40

*With pressure differential of 900 mbar.
Note: Adapted from data of Schleicher & Schuell & Co.

flow rates and a high loading capacity for filtration of very fine particulate matter. From the data in Table 1, it is quite clear that flow rates increase as the pore size increases.

The first two membrane filters in the table are useful for aqueous solutions; the next three can be used with both aqueous and organic solutions. In addition, polycarbonate membranes with pore sizes from 0.2 to 10 μm can be used with the aqueous and organic solutions. PTFE with polyester support is useful for organic solutions, concentrated acids, and bases. These filters are classified as surface filters because the filter matrix acts as a screen and almost entirely retains particulates on the smooth membrane surface. The retention levels for these filters allow efficient retention of sub-micron size particles and organisms. A number of interesting techniques have evolved on the basis of separations on membranes (see Section V. B). Water microbiology and air pollution analyses are some of the major applications of membrane filters.

V. MEMBRANE SEPARATIONS

Permeable barrier separation methods were discussed briefly in Chapter 1, and a list of these methods was provided. It should be noted that most permeable barrier separation processes of practical importance utilize semipermeable membranes as the restrictive interface.[1] These membranes allow passage of certain chemical species completely, while preventing or strongly retarding the permeation of others. Synthetic membranes have been developed that can be used successfully for a number of applications listed here.

- Produce drinkable water from salt water
- Remove urea from blood
- Separate azeotropic mixtures such as alcohol and water
- Recover helium from natural gas
- Remove sulfur dioxide from stack gases

Membranes have also been used as separators in batteries and fuel cells, and as ion-selective tools in analytical chemistry. The familiar separation methods referred to as ultrafiltration, dialysis, electrodialysis, and reverse osmosis (see Section V. B) are based on membrane separations; they have gained increasing attention because they are convenient methods for purification and separation of molecular solutions. Selective membranes have also been developed for the separations of chiral compounds (see Chapter 11, Section III. E).

A. Structure and Properties of Various Membranes

As mentioned previously, a semipermeable membrane allows movement of particular chemical species completely while stopping or strongly restricting the permeation of others. At the same time, the transport rate of permeation has to be high enough to help achieve reasonably rapid separations. Good mechanical and chemical stability are important in these membranes. A brief description of various semipermeable membranes is given below.

I. Microporous Membranes

The structure and function of a microporous membrane is very similar to that of a traditional filter. It has a stiff structure with a great number of voids that are made by randomly distributed interconnected pores of extremely small size (10 to 1000 Å). For the sake of comparison, it should be noted that conventional filters have a pore size greater than 10,000 Å. The membrane rejects the particles or molecules larger than the largest pores, and particles smaller than the smallest pore size pass through the membrane. The particles of intermediate pore size are partially rejected in relation to the membrane's pore size distribution. This suggests that with a microporous membrane, only particles that differ significantly in size can be separated.

2. Homogeneous Membranes

A homogeneous membrane is basically a homogeneous film or interphase through which a mixture of chemical species can be carried by molecular diffusion. The separation of various components in the mixture is related to their transport rate within the interphase; their transportation rates are determined by the diffusivity and concentration of individual components in the film. Particles of exactly the same size can be separated with homogeneous diffusive-type membranes when their solubilities in the film or their concentrations differ significantly.

3. Charged Membranes

Electrically charged membranes are commonly porous in nature. The porous walls carry fixed ions, which may be positively or negatively charged. When positive ions are fixed on the membrane, it is called an anion-exchange membrane. On the other hand, when negatively charged ions are fixed, it is called a cation-exchange membrane. Separation in an electrically charged membrane is accomplished by pore size as well as exclusion of coions (ions of the same charge as the fixed ions) from the membrane.

4. Thin Membranes

Transport rates through membranes are inversely proportional to the thickness of the membrane. In an effort to achieve high transport rates, it is desirable to build membranes that are as thin as possible. For membrane separation processes, such as reverse osmosis or ultrafiltration, significantly thin membranes are essential. This fact has led to a novel membrane fabrication procedure that results in a structure consisting of a thin "skin" (0.1- to 1-μ thickness), which represents the discriminating membrane supported on a 0.1- to 1-mm thick microporous structure. The skin in such a membrane constitutes a porous or homogeneous membrane and may be charged or neutral.

B. Membrane-Based Separation Methods

Numerous applications have been found for membrane separations. A few of these are briefly described here.

1. Ultrafiltration

The term ultrafiltration is commonly used to describe a separation involving solutes of dimensions larger than 20 Å in diameter. In ultrafiltration, a hydrostatic pressure of 1 to 10 atm is applied. Ultrafiltration encompasses all membrane-moderated, pressure-activated separations involving solutions of medium molecular weight (approximately 1000 or greater) solutes, macromolecules, and colloids.

2. Reverse Osmosis

Separation of particles whose molecular dimensions are of the same order of magnitude as those of the solvent is brought about by reverse osmosis. In reverse osmosis, 10 to 100 atm hydrostatic pressure is applied. The technique is used for solutes with a molecular weight up to 500. Most reverse osmosis processes utilize a homogeneous membrane or skin-type membrane. A solution of particles that have a molecular weight of 500 or less may have a notable osmotic pressure that can be as much as 100 atm, depending on the concentration of the solution. In order to achieve successful separation, the osmotic pressure difference between the feed solution and the filtrate has to be overcome by the applied hydrostatic pressure.

The large-scale purification of saline water sources is the most common application of reverse osmosis. It may not be the most economical process for seawater desalination; however, it can be very useful for brackish water purification. Reverse osmosis can be useful for tertiary treatment of sanitary sewage, concentration of whey streams from the milk and cheese industry, and purifying acidic mine waters.

3. Dialysis

A dialyzer is an apparatus in which one or more solutes are transported from one fluid to another through a membrane under a concentration driving force. Two independent factors determine the overall efficiency of a dialyzer: the ratio of flow rate of two fluids and the rate constant of the solute transport between the fluids. The latter is controlled by the characteristics of the membrane, the membrane area, the local fluid velocities, and the fluid channel geometry. An efficient artificial kidney is an excellent example of the application of dialysis. Low molecular weight toxins in the patient's blood, such as urea, creatinine, and uric acid, move across the membrane into a dialyzate solution having a composition such that the osmotic pressure is the same as that of blood, thereby controlling the rate of transport of certain salts. The "cleaned" blood is then returned to the patient.

4. Electrodialysis

In electrodialysis, electrically charged membranes are used to separate components of an ionic solution under the driving force of an electric current. This process has been used for desalination of water, recovery of salt from seawater, de-ashing of sugar solutions, and deacidification of citrus juices.

VI. DISTILLATION

The origin of distillation can be related to evaporation. Distillation can be distinguished from evaporation in that distillation is the separation of a mixture in which all the components of interest are volatile, whereas in evaporation volatile components are separated from nonvolatile ones. Most of these volatile components separated by distillation are generally liquids; however, distillation is not necessarily limited to volatile liquid components. At times, it is possible to volatilize a solid component directly by sublimation, which can be described as a highly specialized distillation method where a solid is directly vaporized without passing through a liquid phase.

A. Phases in Distillation

Distillation is used mainly for separation of the components of liquid mixtures, and it depends on the distribution of constituents between the liquid mixture and vapor in equilibrium with the mixture. The two phases exist by formation of the vapor phase through partial evaporation of the liquid mixture. Each phase can be recovered separately, with the more volatile components concentrated in the vapor, while the less volatile ones are in greater concentration in the liquid.

Various forms of distillation such as fractional distillation, flash distillation, vacuum distillation, steam distillation, and azeotropic distillation are briefly described below.

B. Modes of Distillation

1. Fractional Distillation

Distillation evolved into fractional distillation, which proved to be a boon to the petroleum industry in its quest to isolate various fractions of petroleum. This method is based on the return of a portion of condensate to the distillation unit under conditions such that this condensate is continuously and countercurrently in contact with the vapors. The liquid portion returned to the distillation unit is referred to as reflux, and the method is called fractional distillation or rectification. This method provides great enrichment

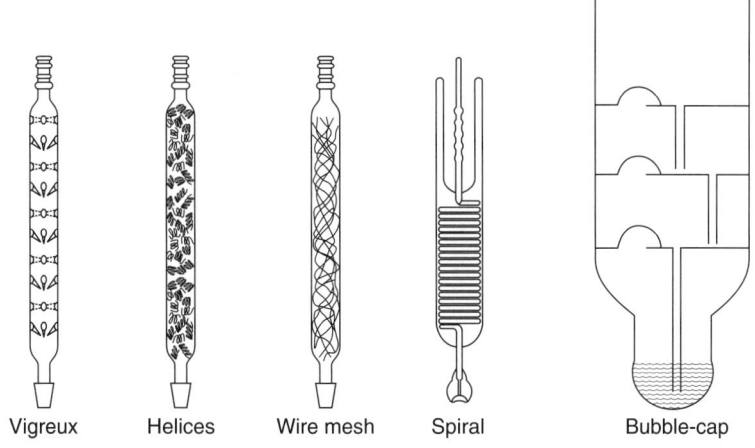

FIGURE 2 Fractionating columns.

of vapors of the more volatile component than produced in simple distillation. It requires the use of multiplate columns, commonly called fractionating columns. Fractional distillation is not restricted to the petroleum industry. It can be useful wherever volatile fractions have to be separated from one another.

Various fractionating columns are shown Figure 2. Fractional distillation, for example, can be carried out with a greater degree of success in a bubble-cap column. The original mixture is heated in a pot to its boiling point. The vapors pass through plate 1 and are deflected and condensed by the bubble cap. Plate 1 is maintained at the boiling point of the mixture at this stage, which is somewhat lower than the boiling point of the original mixture in the pot. The vapors formed at plate 1 are condensed at plate 2, and this process continues to the end of the column, where a condenser is placed. The vapors at each plate are progressively richer in the more volatile component. This means that if we have adequate plates, we can separate any two volatile components in a mixture to the required degree of purity. It may be desirable to mention here that the concept of using plates to evaluate efficiency of a column has been carried into chromatography.

2. Flash Distillation

Flash distillation consists of instantaneous and continuous vaporization of a definite fraction of the liquid mixture in such a way that the total vapor produced is in equilibrium with the residual liquid.

3. Vacuum Distillation

Distillation under decreased pressure is called vacuum distillation and is used to separate high-boiling mixtures or materials that decompose below their normal boiling points. Low pressure reduces the boiling temperatures.

4. Steam Distillation

Steam distillation is utilized to separate high-boiling mixtures or to separate a material from a nonvolatile impurity. The boiling temperature of the mixture is reduced by vaporizing it into a stream of carrier vapor (steam), which, upon condensation, is immiscible with the original mixture and thus can be easily separated from it.

5. Azeotropic/Extractive Distillation

Azeotropic/extractive distillation methods are used for mixtures that are difficult to separate. The relative volatility of the components in the mixture is altered by adding another substance. These methods are useful in the separation of mixtures whose components boil too close together for economic fractional distillation.

C. Distillation as the Precursor to Gas Chromatographic Methods

Fractional distillation can be considered a precursor to the development of gas chromatographic methods, where volatile components are separated on a packed or capillary column. The need to separate nonvolatile components led to the discovery of liquid chromatographic methods such as HPLC that are based on partition (see Section VII). The discovery of newer methods, described in Chapter 11, Section II, may be largely due to our efforts to continually search for more efficient and selective methods. The best example of selectivity is demonstrated by separation of chiral compounds that have the same molecular structure, as well as the same physicochemical properties. These methods are described briefly in Chapter 11 to assist with method selection between simple methods or more sophisticated methods.

VII. EXTRACTION

Extraction is a relatively simple type of separation process in which a solute is distributed between two immiscible solvents. Basic discussion on this topic is available in several standard texts.[1,2]

A. Distribution Coefficient

The ratio of solute in terms of concentration (C) in the two solvents, 1 and 2, is given by the following equation:

$$K_C = C_1/C_2 \qquad (1)$$

where K_C = distribution coefficient, C_1 = concentration of solute in solvent 1, and C_2 = concentration of solute in solvent 2.

To obtain more precise information, activity of each solute instead of concentration should be used in the above equation.

It may be worthwhile to note here that the terms *distribution coefficient* and *partition coefficient* are used interchangeably. An important point to remember is that the distribution coefficient is related to relative solubilities of the solute in the two solvents. Frequently, one of the solvents is aqueous and the other solvent is organic, and it is essential that the selected organic solvent is immiscible with the aqueous solution. When a sample is subjected to the extraction process by shaking it with the two solvents and then allowing them to separate, it is likely that organic ionic species and polar organic compounds will be found to a great extent in the aqueous phase, while the nonpolar organic compounds will remain largely in the organic phase. The familiar expression "like dissolves like" is generally applicable here. In dilute solutions, to a first approximation, the distribution coefficient is independent of concentration.

In Eq. (1) the organic phase is generally placed in the numerator. At times, the lighter phase is placed in the numerator regardless of whether it is organic or not. Thus, the assignment of phase number is arbitrary and should be explicitly stated when quoting distribution coefficients.

B. Distribution Coefficient of a Species

It is important to note that distribution coefficient relates to a single species and should not include possible products of side reactions.[1] The classical example of extraction of benzoic acid, HB, from aqueous solution (whether or not acidified with a stronger acid such as hydrochloric acid to suppress dissociation of benzoic acid) into an organic solvent such as diethyl ether can be used to illustrate this point. When the aqueous solution of benzoic acid is acidified to suppress the dissociation of the acid, the distribution coefficient is given by the following equation:

$$K_C = \frac{[HB]_e}{[HB]_a} \quad (2)$$

where $[HB]_e$ = concentration of benzoic acid in the ether phase, $[HB]_a$ = concentration of benzoic acid in the aqueous phase.

If the aqueous layer is not acidified, benzoic acid dissociates as follows:

$$HB \longleftrightarrow H^+ + B^- \quad (3)$$

and

$$K_a = \frac{[H^+][B^-]}{[HB]} \quad (4)$$

Two independent equilibria are in operation here:

1. Undissociated benzoic acid in the ether layer
2. Dissociated benzoic acid in the aqueous layer

The partition equilibrium relates only to undissociated molecules in the two phases, and the dissociation equilibrium relates only to the species in the aqueous phase, since benzoic acid does not dissociate in the ether phase.

C. Distribution Ratio

A complication can arise if the organic extraction solvent allows dimerization of the solute, such as benzoic acid. For example, if benzene is used as the extraction solvent instead of ether, benzoic acid (HB) would partially dimerize in the benzene phase:

$$2HB \leftrightarrow HB \cdot HB \qquad (5)$$

and

$$K_d = \frac{[HB \cdot HB]}{[HB]^2} \qquad (6)$$

where K_d relates to dimerization of benzoic acid.

An expression that takes all forms into account is called distribution ratio, D. This allows us to determine how much benzoic acid, regardless of its form, is present in each phase, even when dimerization has occurred in one of the phases.

$$D = \frac{\text{total concentration of benzoic acid in organic phase}}{\text{total concentration of benzoic acid in aqueous phase}} \qquad (7)$$

$$= \frac{[HB]_0 + 2[HB \cdot HB]_0}{[HB]_a + [B^-]_a} \qquad (8)$$

From Eq. (4), we can derive

$$[B^-] = K_a \frac{[HB]}{[H^+]} \qquad (9)$$

From Eq. (6), we can get

$$[HB \cdot HB] = K_d [HB]^2 \qquad (10)$$

Substituting these derivations into Eq. (8) gives

$$D = \frac{[HB]_0 + 2 K_d [HB]_0^2}{[HB]_a + K_a [HB]_a/[H^+]_a} = \frac{K_C(1 + 2K_d[HB])}{1 + K_a/[H^+]} \qquad (11)$$

Equation (11) shows that the distribution ratio can be changed simply by changing the pH of the aqueous solution. In acidic solutions, at low pH or high hydrogen ion concentration, D is large and benzoic acid will favor the organic layer. In alkaline solutions, at low hydrogen ion concentration or high pH, D is small and benzoic acid will favor the aqueous layer, largely as benzoate ions.

D. Successive Extractions

When the distribution coefficient is very large and has a value of over 1000, a single extraction can remove essentially all of a solute from the less favorable phase.[2] However, the distribution coefficients are generally not that high for solutes of interest. In that case, it is more effective to divide a given volume of extraction solvent into smaller portions and to use each portion successively

2 SIMPLE SEPARATION METHODS

rather than making one single extraction with the whole volume (see Example 1 below).

I. Double Extraction

To emphasize the importance of successive extractions, let us review an example of double extraction:

Example 1: Calculate the distribution of 8.0 g of a solute from 500 mL of water with 500 mL of diethyl ether (the distribution coefficient of the solute in this system is 3.0 at ambient temperature).

Concentrations After Single Extraction

For a single extraction with 500 mL (0.5 liter) of ether:

$$K_C = C_e/C_a = 3.0 = \frac{(8.0 - x)/0.5}{x/0.5} \tag{12}$$

where C_e and C_a are concentrations in the ether and aqueous phase, respectively, and x is the weight of the solute remaining in the aqueous layer.

From the above calculations, the calculated value of x is 2.0; that is, 2.0 g of the solute would have remained in the aqueous layer and 6.0 g would have been extracted into the ether layer.

Concentrations After Double Extraction

For double extraction, with 250-ml portions of ether

$$K_C = 3.0 = \frac{(8.0 - x_1)/0.25}{x_1/0.5} \tag{13}$$

Calculating for x_1 indicates that 3.20 g of the solute will remain in the aqueous layer after the first extraction, or 4.80 g will be found in the ether layer. The ether layer is then removed, and the second extraction is carried out from the aqueous layer with the remaining 250 ml of ether.

A similar calculation, where x_2 represents the solute remaining in the aqueous layer and 3.2 replaces 8.0 in the above equation, gives a value of 1.28 g of solute remaining after a double extraction, meaning that a total of 6.72 g will be extracted. The results show 12% improvement when double extraction is used (6.72 g vs. 6.0 g).

If five extractions are carried out with 100-ml portions of ether, a similar calculation will show that 7.54 g will be extracted into the ether layer. The results show diminishing returns with increasing numbers of extractions after the fifth extraction.

In general, three extractions are considered optimal for most analytical work. Therefore, an effort is made to select the optimal extraction solvent that is likely to give a high K_C value (see Table 2).

E. Extraction of Mixtures

The extraction of two or more substances in a mixture is of great interest to separation chemists. In these extractions, their distribution coefficients play

TABLE 2 Some Useful Extraction Solvents

Solvent	Boiling point (°C)	Dielectric constant (debye units)
Acetone	56	2.4
Carbon tetrachloride	76	2.2
Chloroform	61	4.8
Cyclohexane	81	2.0
Dioxane	102	2.2
Ethanol	78	24.3
Toluene	111	2.3
Water	100	78.5

a significant role. If one of the two substances has a K_C value much greater than 1 and the other much less than 1 for a given extraction solvent, then a single extraction will bring about nearly complete separation. This is likely to happen if the two solutes are significantly different chemically. On the other hand, if the two solutes have somewhat different distribution coefficients, a single extraction will cause only partial separation, with enrichment of one solute in one phase and the other solute in the other phase. To obtain satisfactory separations, we will have to reextract each phase with a fresh portion of the solvent. A satisfactory separation may be possible with systematic recycling of various intermediate fractions. An apparatus to carry out this process semiautomatically was developed by Lyman Craig.[3] It is described, in some detail, in the next section (Section VII. E).

The mathematical relations describing this process are likely to be helpful in many column chromatographic separations because the distribution profile of a substance as it passes through this apparatus approximates that obtained in a chromatographic separation.

F. Multiple Extractions

To carry out multiple extractions semiautomatically, Craig designed an apparatus that consists of a series of containers connected in a way that allows the outlet of one container to flow into the inlet of the next. Each vessel contains two chambers connected to each other as shown in Figure 3.

The extraction process is initiated by introducing through inlet A an amount of the heavier solvent that will roughly half-fill chamber B.[2] Each of the containers in the train is filled similarly. The solute to be separated is introduced as a solution in the lighter solvent into chamber B of the first container. The assembly is rocked forward and backward at an approximate angle of 35 degrees around a pivoting point, Z. After equilibration is accomplished and the solvents have been separated into two layers, the assembly is rotated 90 degrees in a clockwise motion. The lighter solvent is allowed to flow through connecting tube C into chamber D, while the heavier solvent remains in the lower part of chamber B. When the assembly is rotated

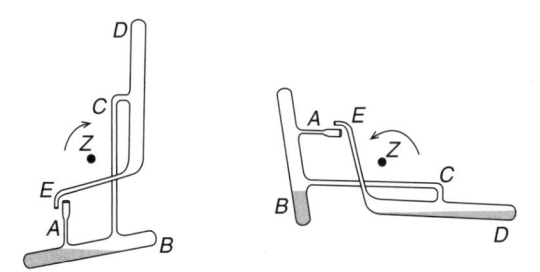

FIGURE 3 Operation of extraction containers of Craig apparatus.

back to its first position, the lighter solvent in D flows through outlet E into chamber B of the next stage. A large number of these assemblies can be operated simultaneously to achieve the desired separation.

To understand the Craig extraction process better, let us assume that each phase occupies half of the volume of the container and K_C of the solute is 1.0. To initiate the operation, the sample is introduced in the first portion of the lighter solvent, L, into container 1. After equilibration, half of the solute is in the upper phase and half is in the lower, heavier phase, H. The upper layer is then transferred to container 2, and a fresh portion of solvent L is added to container 1. After equilibration, one-fourth of the solute is now found in each phase of containers 1 and 2. The solvent L in containers 1 and 2 is transferred to containers 2 and 3, respectively, and fresh solvent L is added to container 1. The distribution of solute by this process can be seen in Figure 4.

The distribution can be represented by binomial $(p + q)^n$, where p is the total solute in phase L of any container, q is the fraction of total solute in phase H of the same container, and n is the number of transfers.

$$(p+q)^n = p^n + np^{n-1}q + \frac{n(n-1)}{2!}p^{n-2}q^2$$
$$+ \frac{n(n-1)(n-2)}{3!}p^{n-3}q^3 + \cdots + q^n \quad (14)$$

The distribution coefficient is given by the following expression:

$$K_C = \frac{C_L}{C_H} \quad \text{and} \quad (15)$$

$$p = \frac{C_L V_L}{C_H V_H + C_L V_L} = \frac{K_C V_H}{K_C V_H + V_L} \quad (16)$$

$$q = \frac{C_L V_L}{C_H V_H + C_L V_L} = \frac{V_L}{K_C V_H + V_L} \quad (17)$$

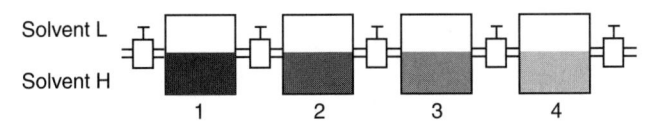

FIGURE 4 Diagrammatic representation of distributions of solute in the Craig extraction process.

TABLE 3 Fraction of Solute in Each Vessel in the Craig Process

No. of transfers	Container number							Factor
	1	2	3	4	5	6	
0	1						$\times 2^{-0}$
1	1	1					$\times 2^{-1}$
2	1	2	1				$\times 2^{-2}$
3	1	3	3	1			$\times 2^{-3}$
4	1	4	6	4	1		$\times 2^{-4}$
5	1	5	10	10	5	1	$\times 2^{-5}$

Note: Adapted from Reference 2.

As per our assumption, $K_C = 1$ and $V_L = V_H$; therefore, p and q are equal to 1/2 as per Eqs. (16) and (17). It is now possible to calculate the amount remaining in any container after a known number of transfers, by carrying out the binomial (Table 3). For example, after five transfers, vessel 1 will contain 1×2^{-5} of the solute and vessel 2 will have 5×2^{-5}, and so on.

The calculations become tiresome when many transfers are involved, but the general approach is direct, by solving the equation $(p+q)^n$ and substituting for the values of p and q as shown in the example given next.

Example 2: Calculate the distribution of a substance after three transfers in a Craig apparatus, for which $V_L = 1$ ml and $V_H = 2$ ml and the distribution coefficient is 2.0.

$$p = \frac{K_C V_L}{K_C V_L + V_H} = \frac{2 \times 1}{2 \times 1 + 2} = \frac{2}{4}$$

$$q = \frac{V_H}{K_C V_L + V_H} = \frac{2}{2 \times 1 + 2} = \frac{2}{4}$$

$$(p+q)^n = \left[\frac{2}{4}\right]^3 + 3\left[\frac{2}{4}\right]^2\left[\frac{2}{4}\right] + 3\left[\frac{2}{4}\right]\left[\frac{2}{4}\right]^2 + \left[\frac{2}{4}\right]^3$$

$$= \frac{8}{64} + \frac{24}{64} + \frac{24}{64} + \frac{8}{64}$$

As a result, the first four containers would contain 12.5%, 37.5%, 37.5%, and 12.5%, of the sample respectively.

Considering that only a single solute is transferred a number of times, we can conclude that the distribution among the containers is obviously a function of several variables: K_C, V_L/V_H, and n.

Distribution of Two Components

Let us now consider a sample that contains two components, A and B, each having a different distribution coefficient. Let us assume $K_A = 0.5$ and $K_B = 1$. If our sample is sufficiently small so that none of the solutions become saturated, each of the solutes will act independently. That is, we can calculate

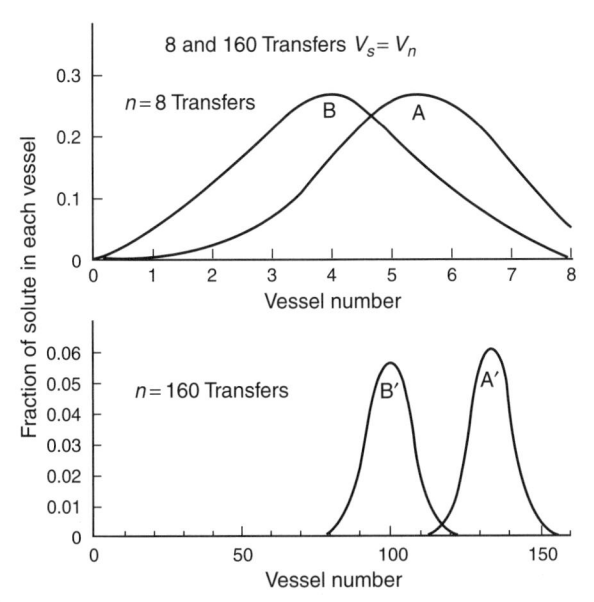

FIGURE 5 Distribution of solutes A and B in the Craig apparatus.[2]

the distribution of each solute as if it were the only one existing. Curve A in Figure 5 shows the portion of total solute A ($K_A = 0.5$) in each container (1 to 9) after eight transfers ($n = 8$). Curve B shows the distribution for solute B ($K_B = 1.0$). If the experiment is carried through 160 transfers, the distributions are as shown in Figure 5, curves A′ and B′.

A number of observations can be made:[2]

1. We see virtually no separation after five transfers with a separation factor ($\alpha = K_B/K_A$) of 2. And eight transfers give only a very partial separation of the solutes. All containers except the first contain notable amounts of solutes A and B.
2. The separation is still incomplete after 160 transfers, but it is probably satisfactory for most purposes. Vessels 80 to 117 contain nearly all of solute B, and vessels 118 to 155 contain nearly all of solute A. To get a higher purity with a lower yield, containers 112 to 123 can be discarded.
3. More transfers would improve the separation but would increase the time and the amount of solvent needed.
4. The concentration of the solutes declines with a greater number of transfers. With eight transfers, the container at the center of the peak (maximum concentration) contains about one-fourth the solute. Even with 200 transfers, the maximum amount of solute in one container will be only about 6% of the total.
5. After eight transfers, the solutes are contained in eight containers. After 200 transfers each solute will be distributed over approximately 50 containers. The "peaks" become broader with a larger number of transfers; however, the solute takes up a smaller fraction of the containers used.

6. The solute is never fully removed from even the initial container, but after the first several transfers, the amount remaining in that container may safely be ignored. For example, if $K_C = 1$ and $V_L = V_H$, then $p = 1/2$, and the amount remaining in the first container is p^n, or 3% ($n = 5$), 0.1% ($n = 10$), 10^{-4}% ($n = 20$).
7. If the separation factor was 1.1 instead of 2, then 10,000 transfers would be necessary to complete the same degree of separation. This is not a very practical technique, even with an automated Craig apparatus.

Chromatography can achieve separations of this type very rapidly and efficiently, and in the next few chapters there will be considerable explanation of these techniques.

G. Countercurrent Extraction Versus Differential Column System

Based on previous discussion, we know that one solvent moves with respect to the other in a discontinuous fashion in the Craig apparatus. To achieve continuous countercurrent extraction, we have to envision that the size of each equilibration vessel is reduced until the entire system becomes a single column and that the extracting solvent can be passed through the system continuously. When both solvents are restricted to thin layers, it is conceivable to approach an equilibrium state at all points in the apparatus, even though it is not exactly completed at any given point. To achieve this goal, one need not use a series of discrete vessels; it is more convenient to construct a column packed with a porous material that will hold one of the solvents stationary on its surface, while permitting the other solvent to pass through it. The stationary solvent need not be immovable; it is necessary only that the two solvents trickle through each other, exposing a large interface. This is the basis for the term *countercurrent extraction*. In this method an extremely large number of equilibrations can be attained without unnecessarily increasing the size of the apparatus.

The relationship of continuous countercurrent extraction to the semiautomatic Craig process is similar to the relationship of continuous fractional distillation with a packed column to the classical bubble-cap plate distillation (see fractional distillation in Section VI). Therefore, it is not uncommon to refer to theoretical plates even in a continuous extraction column. It should be realized that this expression refers to the number of distinct equilibrations and transfers that would have to be done in order to accomplish an equivalent level of separation.

I. Relating Countercurrent Extraction to Partition Chromatography

Countercurrent extraction is similar to partition chromatography, where one of the phases remains fixed on the column and is called the stationary phase, while the other percolates through the column and is called the mobile phase.

A number of equilibrium separation techniques can be carried out in either stage or differential column (a packed-bed column) operations. If discrete stages, for example, plates, are used, the process can be modeled as a collection of equilibrium stage processes, although in practice, complete equilibrium is not reached at each stage. The efficiency of the separation equipment is often expressed in terms of the number of theoretical stages or plates.

Separation processes in a differential column system are more complicated because changes in composition are continuous, and hydrodynamic effects and mass transfer in both phases have a great influence on the rate of composition change. Although a differential column has no stages, the theoretical stage concept is still applicable and leads to a convenient measure of efficiency. From the results obtained with a differential column, the number of theoretical plates that characterizes the efficiency of the device can be calculated by comparison with a hypothetical device as discussed above. Often, the length of the column is divided by the number of theoretical plates. This length is termed the height equivalent of a transfer unit (HETU) or the height equivalent of a theoretical plate (HETP). Obviously, larger plate height suggests lower efficiency of the separating device. In other words, a more efficient column requires a shorter column segment to achieve the same degree of separation, that is, theoretical plate height should be smaller. Although in most cases sophisticated rate theories that account for various phenomena taking place in a differential segment of the column are available, the theoretical plate model remains a useful and convenient approach to describe and design not only stage, but also differential column processes.

REFERENCES

1. Karger, B. L., Snyder, L. R., and Horvath, C. *An Introduction to Separation Science*, Wiley, New York, NY, 1973.
2. Pecosk, R. L., Shields, L. D., Cairns, T., and McWilliam, I. G. *Modern Methods of Chemical Analysis*, Wiley, New York, NY, 1976.
3. Craig, L. C. and Craig, D. In *Technique of Organic Chemistry* (A. Weisberger, Ed.), Vol. III, p. 150, Wiley, New York, NY, 1956.

QUESTIONS FOR REVIEW

1. Define distribution coefficient.
2. Two grams of benzoic acid are dissolved in 200 ml of water and extracted with 200 ml of diethyl ether. The distribution coefficient of benzoic acid is 100, and its dissociation constant is 6.5×10^{-5}. Calculate the distribution ratio (D) of benzoic acid at pH 2, 5, and 6.
3. Calculate D at pH 2 to 10 (1 unit apart) in the above problem, and plot D versus pH.
4. Describe the operation of the Craig apparatus.

3 EQUILIBRIUM PROCESSES IN SEPARATIONS

I. MOLECULAR INTERACTIONS
II. SEPARATION THERMODYNAMICS
   A. Separations Based on Phase Equilibria
   B. Distribution Equilibria–Based Separations
REFERENCES
QUESTIONS FOR REVIEW

Based on discussions in Chapters 1 and 2, we know that a desired separation can be obtained by the transferring chemical components from one region to another. Since samples requiring separation contain two or more components, this necessitates a change in the relative concentration of two or more components within a defined region. As pointed out in Chapter 1, either rate or equilibrium processes can be involved in the transfer of a given component. The methods based on rate processes relating to permeable barrier separations such as membrane separations were covered in Chapter 2; other discussion of interest may be found in Chapters 5 and 11 of this book. We will review equilibrium processes as they relate to separations in this chapter (the molecular basis of separations is covered in Chapter 4).

The separation systems at equilibrium need to be heterogeneous (the notable exceptions are molecular exclusion processes), involving extraction, partition, or transfer of components between two or more distinct, generally immiscible phases (see Chapter 2). It is possible to achieve a separation based on equilibrium if there is a difference in the distribution of components between the two immiscible phases.[1] At this stage, it may be a good idea to review what we mean by the terms *phase, phase equilibria,* and *distribution equilibria*.

Phase

A phase is a definite portion of a given system that is homogeneous throughout and is physically separated from the other phases by distinct boundaries.

Phase Equilibria

The phases are actually composed of the components we want to separate on the basis of phase equilibria. An excellent example is separation of a two-component mixture by distillation.

Distribution Equilibria

The distribution equilibria separations are based on equilibria where there is distribution of components between the two immiscible phases. At times, the phases may be composed of chemical species that are not part of the mixture we want to separate.

When only two phases are involved, the separations that are dependent on phase or distribution equilibria can be classified as follows, on the basis of the type of phases involved:

Phases	*Examples of Separations*
Liquid–solid	Precipitation; adsorption, liquid–solid adsorption chromatography
Liquid–liquid	Partition; liquid chromatography, HPLC
Gas–solid	Sublimation; adsorption, gas–solid chromatography
Gas–liquid	Distillation; partition, gas–liquid chromatography

Adsorption equilibria, for example, are influenced by the surface of the bulk phase. Notable examples are gas–solid or liquid–solid adsorption chromatography. It may be instructive to note that adsorption can play some role in various separation methods, including modern liquid chromatography, even when partition or ion exchange is the primary mode of separation. The interpretation and prediction of separation equilibrium can be based on two approaches: molecular interactions and thermodynamics.

I. MOLECULAR INTERACTIONS

The interactions of physical and chemical forces between molecules within each phase are called molecular interactions. The equilibrium distribution is determined by these interactions, which in turn can be related to the molecular structure of the interacting molecules. This suggests that it should be possible to estimate phase and distribution equilibria from a knowledge of the structure and concentration of the compounds comprising an equilibrium system (see further discussion in Chapter 4).

Models based on molecular structure allow more detailed predictions than are possible with thermodynamics, but these models tend to be less accurate because only minimum information is generally available about the sample. Hence, these models can be applied to only a small range of separation systems. Because of these limitations, a greater reliance has been placed on thermodynamics when interpretation or prediction of separation equilibria must be made (see the discussion that follows).

II. SEPARATION THERMODYNAMICS

Based on our preliminary knowledge of physical chemistry, we know that thermodynamics is the study of quantitative relations between heat and other forms of energy (it may be desirable at this stage to review your physical chemistry text on that subject). The important fundamental thermodynamic principles in separation will be reviewed here. The distribution equilibria for separation of components at low concentrations will be used to achieve this objective.

Now let us consider the distribution of a solute, s, between two immiscible phases, x and y, at constant temperature and pressure:

$$s^x \leftrightarrow s^y \qquad (1)$$

The separation of a solute is determined by the ratio of solute (sample component of interest) concentration in the two phases, which can be defined by an equilibrium concentration distribution coefficient, K_x, when mole fractions of the solute are used in the calculations.

$$K_x = X^x/X^y \qquad (2)$$

where X^x = mole fraction of s in phase x abd X^y = mole fraction of s in phase y.

When molar concentration terms are used in the equation instead of mole fractions, this equation takes the following form:

$$K_c = C^x/C^y \qquad (3)$$

Since $C = X/V$ for dilute solutions, the above-mentioned distribution coefficients can be related as follows:

$$K_c = K_x(V^y/V^x) \qquad (4)$$

where V^y = molar volume of phase y and V^x = molar volume of phase x.

The chemical potential μ (partial molar free energy) of the solute, s, in a given phase is given by the following equation:

$$\mu = (\partial G/\partial n_s)_{T,P} \qquad (5)$$

where G = the free energy of the total phase and n_s = the number of moles of s present in the phase.

At equilibrium the net change in free energy for transfer of solute, s, between phases x and y must be zero, so that the chemical potentials for s in each phase must be equal:

$$\mu^x = \mu^y \qquad (6)$$

The value of μ in any phase can be expressed as follows:

$$\mu = \mu^0 + RT \ln a \qquad (7)$$

where a = the activity of the solute in a particular phase and μ^0 = the chemical potential of the solute in a chosen standard state.

Since the standard state can be described only by its temperature, composition, and pressure, the quantity μ^0 is a constant, and $a = 1$ for the solute in the standard state; that is, μ is equivalent to μ^0.

A. Separations Based on Phase Equilibria

As mentioned earlier, phase equilibrium separations differ from those based on distribution equilibria in that they involve phases that are composed of the mixture to be separated. Such separations are useful in preparative- and production-scale applications because no species that require subsequent removal are added in the process. Phase equilibrium separations are also more suitable for relatively simple mixtures containing only a few components of interest. Let us review phase rule and certain basic aspects of one- and two-phase equilibrium systems.

1. Phase Rule

In heterogeneous systems the composition and physical properties are different in different parts of the systems. A well-known example of a heterogeneous system containing three phases is a system containing water, ice, and water vapor. The conditions under which different phases can exist may be conveniently described by phase diagrams. Time is not considered a variable here because phase diagrams are used for systems that are in equilibrium.

It is important to recall at this point that the phase rule allows us to interpret phase diagrams based on sound thermodynamics. The phase rule was discovered in 1876 by Professor J. Willard Gibbs of Yale University and is called Gibbs phase rule, or, simply, the phase rule. To develop a better understanding of phase equilibria, let us review the Gibbs phase rule, which states

$$P + F = C + 2 \qquad (8)$$

where P = number of phases, F = degrees of freedom, and C = number of components.

Simply stated, this rule means that for a given number of components C, in a system, we can predict the number of phases P, that can coexist for a specified number of degrees of freedom F. The degrees of freedom compare to the independent intensive variables of the system (temperature, pressure, or the concentration of the components in the given phases). Similarly, if we know C and P, we can ascertain the number of variables F that can vary independently of one another.

2. Single-Component Systems

A phase diagram is used to describe a phase equilibrium system and to utilize the system to achieve separations. Figure 1 shows phase diagrams for a single-component system in terms of two variables: temperature T and pressure P.

Based on the phase rule equation given above, we can calculate that a single component in one-phase system requires that two degrees of freedom (e.g., temperature and pressure) be specified to define the system. For example, point a in Figure 1 represents a pressure and temperature at which only the gas phase exists. Point b corresponds to a pressure and temperature at which two phases, gas and liquid, coexist in equilibrium with each other. Point c is

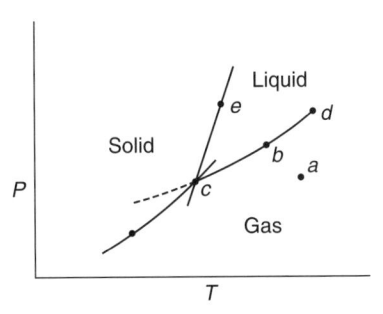

FIGURE 1 A typical phase diagram for a single component.

the *triple point*, at which all three phases are in equilibrium. Since $F=0$, the triple point is unique, existing at only one pressure and temperature.

In Figure 1 the gas–liquid line ends at the *critical point* (point d), corresponding to the critical temperature, T_c, and critical pressure, P_c, of the sample. The attributes of the gas and the liquid become of equal value at the critical point. Regardless of the pressure level, at temperatures higher than the critical temperature only one phase exists (the so-called supercritical fluid). It may be worthwhile to mention at this point that the supercritical fluids are very useful for carrying out difficult extractions. They are also used in chromatographic separations in a technique called supercritical fluid chromatography (this topic is covered in Chapter 11).

3. Two-Component Systems

For two-component systems, the phase diagrams become more complicated. If only one phase is present, then $F=3$, from the phase rule; and the three independent variables are P, T, and the mole fraction of one of the components (the mole fraction of the other component is obtained by the difference). We need to know only two variables when two phases are in equilibrium. Here it is convenient to use the mole fraction of one component in one of the phases, and either the temperature or the pressure as the independent variable.

To illustrate the usual type of phase equilibrium for a lower-boiling component, l, and a higher-boiling component, h, three possible types of phase diagrams for vapor–liquid equilibrium in a two-component system are shown in Figure 2.

In Figure 2A, let us look at point d in the two-phase region of liquid and vapor, corresponding to temperature T_1, and composition (total system) $X_i = c$. At T_1, the vapor of composition a, is seen to be in equilibrium with liquid of composition, b. The horizontal line through point d is called the *tie line*. It should be noted that the vapor phase is enriched in the more volatile component l, and the liquid phase in the less volatile component h. As a result, partial separation occurs, and this provides the basis for distillation.

Figures 2B and C present the circumstances of an azeotropic, or constant-boiling, solution. In the figure the point marked S corresponds to maximum or minimum boiling temperature. The composition of both vapors and liquids is the same, and the composition of the liquid stays constant while increasing amounts of liquid are converted into vapors. A liquid composition, S

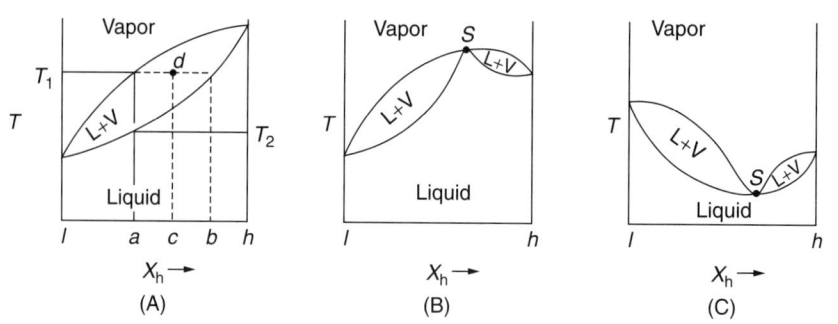

FIGURE 2 Phase diagrams of a two-component system.

acts like a pure compound and cannot be separated by distillation using the pressure of the phase diagram. However, in some situations three-component azeotropic solutions are purposely created to amplify the separation of a two-component mixture using distillation.

A type of phase diagram commonly observed in liquid–solid phase equilibria for two-component mixtures is shown in Figure 3. As was shown earlier, the temperature is plotted against composition at constant pressure. The two liquids are completely miscible, and solid solutions are not formed (it is important to recognize that in reality, zero solid solubility probably never occurs). For example, at point u we can calculate the relative amounts of the liquid and solid phases.

By cooling the mixture (composition c) below T_1, more solid, l, forms, and the liquid solution composition follows the phase equilibrium line to point e, the eutectic point; and three phases coexist at equilibrium at the eutectic point, so that $F = 1$ from the phase rule. The calculated one degree of freedom is specified by the given pressure dictated by the phase diagram; that is, the *eutectic point* does not change at constant pressure.

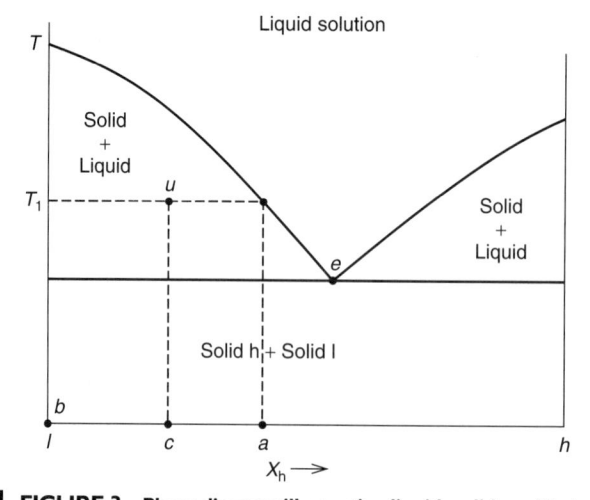

FIGURE 3 Phase diagram illustrating liquid–solid equilibrium.

B. Distribution Equilibria–Based Separations

As previously mentioned, the two phases in distribution equilibria–based separations are composed of compounds that are different from the sample components. These systems are generally interesting for analytical or preparative separations because of the following reasons:

1. The combined volume of the two phases is much greater than that of the sample and can be unwieldy on a production scale.
2. The distribution coefficient is usually constant.

Separations based on distribution equilibria have certain advantages over corresponding separations by phase equilibria. The composition of one or both phases in distribution equilibria may be varied over broad limits. This in turn has an effect on the distribution constant (K) values of individual sample components and thus allows more difficult separations to be accomplished.

I. Distribution Isotherms

To better understand distribution isotherms, let us first review adsorption of gases by solids. The basic theory of adsorption of gases by solids was developed by Langmuir early in the twentieth century. He considered the surface of a solid to be made up of elementary spaces, each of which could adsorb one gas molecule. It was assumed that all spaces were identical and that the presence of a gas molecule did not affect the properties of neighboring spaces. Then, at adsorption equilibrium the rate of evaporation of the adsorbed gas is equal to the rate of condensation.

The relationship between the amount of gas physically adsorbed onto a solid and the equilibrium pressure or concentration at constant temperature, when plotted, provides an adsorption isotherm. These isotherms may be linear, concave, or convex (Fig. 4). Other types of isotherms, such as sigmoidal, are also possible when gases undergo physical adsorption onto nonporous solids to form monolayers followed by multilayers.

In the case of adsorption of solutes—usually onto a solid phase—there is frequently competition among solute molecules for a fixed amount of adsorption sites. In this situation isotherm nonlinearity is not inevitably the consequence of changing intermolecular interactions, but it is essentially statistical in nature.

Let us derive the Langmuir isotherm to illustrate this point and to see under what conditions it is linear. Adsorption can be looked upon as the reversible

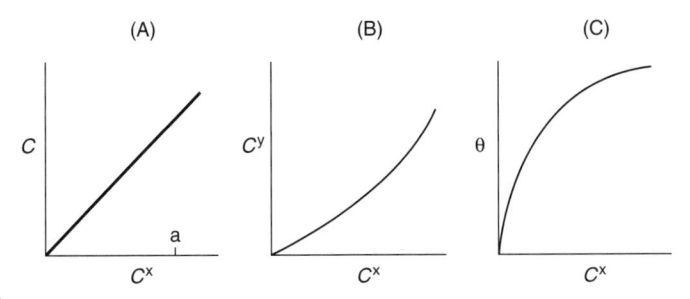

FIGURE 4 Typical distribution isotherms.

reaction of solute, such as a gas, G, in the nonadsorbed (gas or liquid) phase with an adsorption site, A, to form adsorbed G, which can be denoted as GA:

$$G + A \leftrightarrow GA \quad (9)$$

A distribution coefficient, K, may be defined as follows for this reaction:

$$K = \frac{X_{GA}}{X_G X_A} \quad (10)$$

where X_{GA} = mole fractions of GA, X_G = mole fractions of G, and X_A = mole fractions of A.

K is constant if we suppose that intermolecular interactions in each phase are constant as X_G and X_{GA} are varied. By defining the fractional coverage of sites by adsorbed solute as θ, then $X_{GA} = \theta$ and $X_A = 1-\theta$. Placement of these quantities into the equilibrium constant expression produces the Langmuir isotherm, as shown in Figure 4C:

$$K = \frac{\theta}{X_G(1-\theta)} \quad (11)$$

$$\theta = \frac{K X_G}{1 + K X_G}$$

In the case of a gas as a nonadsorbed phase, P_i replaces X_G. At low-solute concentrations, $\theta = K X_G$, and we will obtain a *linear isotherm*. For large values of X_G, θ approaches a limiting value of one, and the distribution coefficient $K = \theta / X_G$ is close to zero.

2. Distribution Equilibria of a Solute

The distribution equilibria for a given solute and system at a particular temperature can be best described by a plot of solute concentration in one phase, x (C^x), versus concentration in the other phase, y(C^y). These distribution isotherms, as shown in Figure 4, are useful because they show how K varies with sample concentration.

Since $K = C^x/C^y$, the value of K for a specific solute concentration is given by the slope of the line through the origin and the corresponding point on the isotherm. If K is constant over some range of solute concentration, the corresponding isotherm is linear through the origin. Over a wide range of solute concentration, true distribution isotherms are rarely linear; they may be either concave or convex (see representations B and C in Fig. 4). However, all isotherms became linear when concentrations are sufficiently dilute, as is demonstrated in Figure 4A. This linear isotherm area, where the distribution coefficient is transformed to the distribution constant, is of considerable interest in analytical separations.

Linear distribution isotherms are comparable to Henry's law for gas–liquid equilibria:

$$\frac{P_i}{X_i} = K^0 \gamma^y \quad (12)$$

$$P_i = C' X_i \quad (13)$$

where P_i is partial pressure of solute in the gas phase, X_i is mole fraction of i in the system, K^0 is thermodynamic distribution constant, γ^y is activity coefficient of solute in phase y and C' is a constant.

Linear distribution isotherms are also equivalent to the Nernst distribution law in the instance of liquid–liquid equilibria:

$$K_x = K^0 \left(\frac{\gamma^y}{\gamma^x}\right) \tag{14}$$

or

$$K = C'' \tag{15}$$

C' and C'' are constants because the activity coefficients, γ, are constant over the solute concentration range where these laws hold, that is, in dilute solutions.

3. Secondary Chemical Equilibria

The introduction of a reversible chemical reaction into a separation system generally leads to secondary chemical equilibria. This affords an additional possibility for the control of the K values of individual sample components. It is a good idea to keep in mind that separation depends in the long run on differences in the K values of individual sample components. The larger differences lead to easier separations.

Clear examples of the use of secondary chemical equilibria to promote distribution separations are provided by acid and base extractions (it may be useful to review the discussion on extraction in Chapter 2). Organic acids or bases may be separated from neutral organic compounds that are preferentially held in the organic phase.

Let us review an example of the extraction of an organic acid, HA, from aqueous medium, a, to an organic solvent, o, to study the effect of secondary chemical equilibrium:

$$(HA)_a \leftrightarrow (HA)_o \tag{16}$$

In the aqueous phase, the acid is likely to dissociate as shown:

$$H_2O + (HA)_a \leftrightarrow (H_3O^-)_a + (A^-)_a \tag{17}$$

The distribution constant as defined by Eq. (16) is

$$K = \frac{[HA]_o}{[HA]_a} \tag{18}$$

The acid-dissociation constant from Eq. (17) may be written as follows:

$$K_a = \frac{[H_3O^+]_a[A^-]_a}{[HA]_a} \tag{19}$$

To simplify, let us assume dilute concentrations are involved, so that K is not dependent on the initial concentration of HA in water.

As discussed in Chapter 2, the distribution ratio, D, deals with the influence of chemical equilibria on the overall distribution:

$$D = \frac{\text{stoichiometric concentration of HA in the organic phase}}{\text{stoichiometric concentration of HA in the aqueous phase}}$$

or

$$D = \frac{[HA]_o}{[HA]_a \cdot [HA]_a} \quad (20)$$

Substituting Eqs. (18) and (19) into Eq. (20), we obtain

$$D = \frac{K[H_3O^+]}{[H_3O^+] + K_a} \quad (21)$$

4. Effect of pH on Extractions

The following example will help illustrate the effect of pH on extractions.[2]

Example 1: 0.2 g of benzoic acid dissolved in 100 mL of aqueous solution is shaken with 100 mL of an immiscible organic solvent. The dissociation constant, K_a, of benzoic acid is 6.5×10^{-5}. Assuming the distribution constant, K, at equilibrium is 100, calculate the distribution ratio, D, if the aqueous layer is at pH 2 and 8.

$$\text{At pH 2:} \quad D = \frac{100}{1 + 6.5 \times 10^{-5}/10^{-2}} = 99.35$$

$$\text{At pH 8:} \quad D = \frac{100}{1 + 6.5 \times 10^{-5}/10^{-8}} = 0.015$$

The results show that 99.35% of the sample would be extracted at pH 2.0 and only 0.015% would be extracted at pH 8.0, which indicates that there is a significant decrease in the extraction of acid as the pH is increased from 2 to 8.

To appreciate the effect of pH on the extraction of benzoic acid, it may be instructive to carry out this process at pH 1 to pH 10 and plot the curve.

Equation (21) can be useful in helping us understand the distribution of the organic acids, exemplified by HA here, over a whole range of acid and base concentrations in water. Let us consider acidic conditions in which $[H_3O^+] \gg K_a$. Then $D = K$ from Eq. (21), and the solution is acidic enough so that no ionization takes place in the aqueous phase and HA acts as a simple neutral species that is disbursed between the two phases.

When a solution pH is such that $[H_3O^+] = K_a$, then $D = K/2$ [Eq. (21)] and half the earlier amount of material is now extracted into the organic phase. The reason for the reduction in amount extracted is that concentration of undissociated HA is only 50% of the first amount of HA added to solution. Because the distribution coefficient must be maintained constant, the concentration of HA in the organic phase must decline by half of that present when no ionization of HA occurs.

In a basic solution where $[H_3O^+] \ll K_a$, Eq. (21) exhibits that $D \ll K$, and very little HA will be extracted. Because of the ionization of acid, the concentration of HA in the aqueous phase is very low, so that the quantity extracted is also quite small. From this discussion it should be clear that the distribution of HA can be greatly influenced by the concentration of the common ion H_3O^+. This allows us to separate different acids with different K_a values. The greater the difference in K_a for two acids, the easier their separation becomes. It should also be noted that in the case of secondary equilibria, the distribution ratio, D, gives a much better picture of the distribution process than do the distribution coefficients.

Another example of the use of secondary equilibria is in the prevention of precipitation by utilizing masking agents. The pH of the electrolyte medium controls the net charge on a given protein. This allows us to separate proteins conveniently by electrophoresis, where species with an overall (net) negative charge migrate toward the positive electrode, and positively charged species migrate toward the negative electrode (see Chapter 11).

5. Capacity Factor

The separation in a distribution system is influenced mainly by the relative amount of solute in each phase, rather than by the relative solute concentrations. For example, if the solute has a K value of 1000, essentially all of it can be extracted if equal volumes of each phase are used. However, slightly more than 99% of the total solute would be found in the phase favored by K if the K value were 100. And if the volume of the preferred phase were to be reduced by a factor of 100, in this case with no change in the volume of the other phase, only 50% of the solute would be extracted in the preferred phase. The fraction of total solute, Φ^x, in a given phase x is given by the following equation:

$$\Phi^x = \frac{C^x V^x}{C^x V^x + C^y V^y} \tag{22}$$

where C^x and C^y = solute concentration in the x and y phases and V^x and V^y = the volume of the phases.

We know that the solute distribution constant K is equal to C^x/C^y; therefore

$$\Phi^x = \frac{K(V^x/V^y)}{1 + K(V^x/V^y)} \tag{23}$$

Capacity factor, k', can be defined as follows:

$$\begin{aligned} k' &= \frac{\text{total amount of solute in phase x}}{\text{total amount of solute in phase y}} \\ &= \frac{C^x V^x}{C^y V^y} \\ &= K(V^x/V^y) \end{aligned} \tag{24}$$

By combining Eqs. (23) and (24), we can calculate the fraction of solute in phase x:

$$\Phi^x = \frac{k'}{1+k'} \tag{25}$$

In phase y, Φ^y is equal to $1/1 + k'$.

At this point it may be worthwhile to point out that the capacity factor, k', is a very useful parameter in chromatography (see Chapter 6).

6. Separation Factor

When two or more solutes are involved, separations can be accomplished only when the K values of the solute are different; that is, there must be some distinction in the thermodynamic behaviors of the components.

Let us define the separation factor α for a pair of solutes e and f:

$$\alpha = \frac{C_e^x C_f^y}{C_e^y C_f^x}$$
$$= K_e / K_f \qquad (26)$$

where C_e^x, C_f^x, C_e^y, and C_f^y pertain to the concentrations solute e or f in phase x or y. K_e and K_f are the corresponding distribution constants.

The separation factor has the same relationship to capacity factors as shown in Eq. (26) for the distribution constants.

We need to emphasize here that high α values are often difficult to obtain for compound pairs that we may want to separate by some distribution process. This problem may be overcome by the use of secondary chemical equilibria in some situations. Acceptable separations based on α values as small as 1.01 may also be achieved, however, using countercurrent procedures, such as Craig distribution (see Chapter 2) or high-resolution gas or liquid chromatography (see Chapters 9 and 10).

It is important to remember that the definition of separation factor for phase equilibria system is basically comparable to the use of α in distribution systems. Also, the separation factor α is a very useful gauge of separation ability of a chromatographic system (see Chapter 6).

REFERENCES

1. Karger, B. L., Snyder, L. R., and Horvath, C. *An Introduction to Separation Science*, Wiley, New York, 1973.
2. Pecosk, R. L., Shields, L. D., Cairns, T., and McWilliam, I. G. *Modern Methods of Chemical Analysis*, Wiley, New York, NY, 1976.

QUESTIONS FOR REVIEW

1. Define phase rule.
2. Derive equations for: (a) Capacity factor (b) Separation factor

4 THE MOLECULAR BASIS OF SEPARATION

I. MOLECULAR INTERACTIONS
   A. Dispersion Forces
   B. Nondispersion Forces
II. SOLUBILITY PARAMETER THEORY
III. GROUP INTERACTIONS
   REFERENCES
   QUESTIONS FOR REVIEW

In this chapter we will review how molecular structures affect distribution of a compound between two phases. This information can be very useful in optimizing the separation of a mixture of known compounds. The same general principles may be applied to the separation of the mixtures of completely unknown compounds, because it is frequently possible to make logical assumptions as to what types of compounds may be present. It goes without saying that background information about a given sample in terms of number of components, the structure of the components, their approximate concentrations, and so forth is very helpful in solving a particular separation problem.

We learned in Chapter 3 that it is possible to calculate the extent of separation in a different system if we have data such as vapor pressure, solubility, etc., for each component as a function of temperature, pressure, and phase composition. Data of these types, however, are frequently not available for a separation system of interest. Therefore, a careful study of intermolecular interactions is helpful in designing a given separation.

I. MOLECULAR INTERACTIONS

It was mentioned in Chapter 3 that interactions of physical and chemical forces between molecules within each phase are called molecular interactions. The lack of strong attractive forces between the molecules allows a gas to consist of widely separated molecules in constant motion. In liquids, the intermolecular forces are strong enough to hold the molecules close together. The discussion here is primarily focused on these interactions between the molecules, or intermolecular interactions. The equilibrium distribution is determined by these interactions, which can be related to the molecular structure of the interacting molecules.[1] This suggests that it should be possible to estimate phase and distribution equilibria from the information of the structures and concentrations of the compounds making up an equilibrium system.

Interaction energy

The distribution of a component **c**, in a given phase can be greatly influenced by partial molar enthalpy, h_c, which can be directly related to the forces between **c** and the molecules surrounding it. In turn, these h_c values affect, to a great extent, the corresponding chemical potentials μ_c and the values of K for the distribution of **c** between different phases. There is a greater tendency for **c** to concentrate in a phase where the intermolecular interaction energy between **c** and the surrounding molecules increases. By calculating approximately the ranges of these interaction energies, we can estimate the relative distribution of any component between two phases.

The result of both attractive and repulsive forces is the net interaction energy, E_{cd}, between two neighboring and nonbonded atoms (or molecules) **c'** and **d'**. Overall,

$$E_{cd} = \frac{A}{r^{12}} - \frac{\sum B}{r^x} \qquad (1)$$

where r is the distance between the nuclei of atoms **c** and **d**; A and B are the constants for particular atoms **c'** and **d'**, $x \leq 6$.

Equation (1) includes one of the major repulsive interactions, namely, the universal tendency of two objects to avoid taking up the same space, and several conceivable attractive interactions. Negative E_{cd} denotes the net attraction between **c'** and **d'**. With coefficient x usually being less than or equal to 6, attractive forces are more meaningful at large interatomic separations and repulsive forces dominate at small values of r. When r falls under a specific value (r_e)—the equilibrium separation distance of **c'** and **d'**—the dependence of repulsive energy on r^{-12} leads to a very abrupt increase in E_{cd} (see Fig. 1).

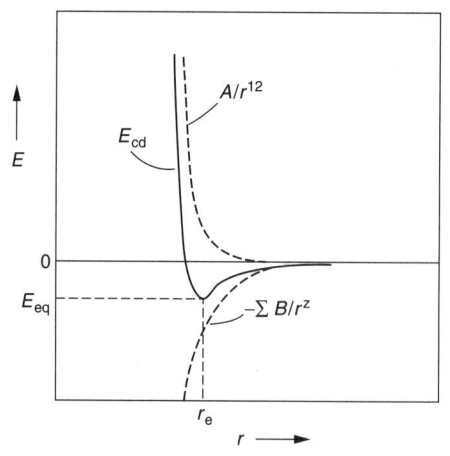

FIGURE 1 Nonbonding interaction energy as a function of separation distance (adapted from Reference 1).

This clearly suggests that the equilibrium separation r_e, or van der Waals separation, is determined primarily by the repulsive term. Also, the net energy of interaction E_{eq} (at equilibrium) may be considered primarily a function of the attractive interactions.

Two types of attractive interactions of particular interest are discussed next; they are influenced by dispersion and nondispersion forces.

A. Dispersion Forces

1. London or Dispersion Force

The weak electrostatic force by which nonpolar molecules such as hydrogen gas or benzene attract one another was first recognized by a well-known scientist named Fritz London in 1930. This force has been described as dispersion or London force.

London force is sufficient to bring about condensation of nonpolar gas molecules so as to form liquids or solids, when the molecules are brought quite close to each other. Such forces exist between any adjoining pair of atoms or molecules. For most organic compounds, dispersion forces account for a greater part of the total attractive energy. In some cases, for example, hydrocarbons, they explain essentially all of E_{cd}. In other situations, such as with rare gases, they are the only possible attractive interaction.

Dispersion forces may be explained as follows: at any given time, various electrons of an atom or molecule may assume any of a number of particular positions through oscillating movements. Each such configuration is identified by some degree of electrical dissymmetry, in that the atom or molecule has an overall dipole or multipole moment. This dipole polarizes the electron clouds in neighboring atoms or molecules, inducing a dipole of opposite polarity, which attracts the original dipole. The attractive interaction energy $(E_{cd})_a$ between two atoms **c** and **d**, which is a result of these dispersion forces may be written as follows:[2]

$$(E_{cd})_a = \frac{-3}{2}\left[\frac{I_c I_d}{I_c + I_d}\right]\left[\frac{\alpha_c \alpha_d}{r^6}\right] \qquad (2)$$

where I_c and I_d refer to the first ionization potentials of atoms c' and d', α_c and α_d are their respective polarizabilities.

Generally, interatomic distances and ionization potentials do not vary much for different pairs of adjoining atoms, especially in the case of organic compounds. Thus, a reasonable approximation would give

$$(I_c + I_d) \approx 2(I_c I_d)^{1/2}$$

and

$$r = (r_c + r_d) \approx 2(r_c r_d)^{1/2}$$

When we use these expressions in Eq. (2), we get

$$(E_{cd})_a = \frac{-3}{8} \left| \frac{I_c \alpha_c^2}{r_c^6} \right|^{1/2} \left| \frac{I_d \alpha_d^2}{r_d^6} \right|^{1/2} \quad (3)$$

The corresponding interaction energies, E_{cc} and E_{dd}, may be defined for interactions between comparable atoms **c** or **d**, respectively. To obtain the value from E_{cc}; I_c, α_c, and r_c are substituted into Eq. (3) for I_d, α_d, and r_d, respectively. Equation (3) may then be presented in the form:

$$(E_{cd})_a = (E_{cc})^{1/2}(E_{dd})^{1/2} \quad (3a)$$

This equation is called the *Geometric Mean Rule* of dispersion interactions.

2. Dispersion Interactions and Heat of Vaporization

The energy needed to move a molecule **c** from the pure liquid into the gas phase is mainly a measure of the interactions between molecules in the liquid phase, since gas-phase interactions are minimal and can usually be ignored. As a result, the heat of vaporization or the boiling point of various compounds may contribute helpful information relating to the size and character of these forces.

Let us now consider the vaporization of nonpolar molecules, whose energy is resolved almost completely by dispersion interactions. A plot of boiling point T_b versus molar volume for several types of hydrocarbons—cycloalkanes, *n*-alkanes, and unsubstituted aromatics—is illustrated in Figure 2. Heats of vaporization, $\triangle H^v$, at the boiling point are also given in the figure.

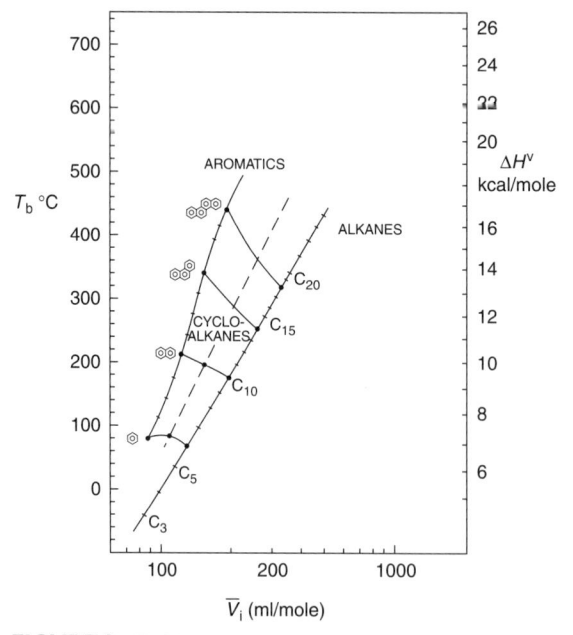

FIGURE 2 Boiling points or heats of vaporization of hydrocarbons versus molar volume (adapted from Reference 1).

4 THE MOLECULAR BASIS OF SEPARATION

TABLE I Effect of Dipole Moment and Hydrogen Bonding on Boiling Points of Some Polar Compounds

	Boiling point (°C)			Dipole moment (D*)
Compound	Experimental	Calculated	Difference	
Hydrocarbons			0	0
Diethyl ether	35	5	30	1.2
Methyl acetate	57	−26	83	1.7
Acetone	56	−31	87	2.8
Acetonitrile	80	−66	146	3.9
o-Dichlorobenzene	179	159	20	2.3
m-Dichlorobenzene	172	159	13	1.5
p-Dichlorobenzene	174	159	15	0.0
Methyl alcohol	65	−86	151	1.6
Water	100	−115	215	1.8

D* = Dipole moment in solution. Note: Adapted from Reference 1.

There is a close parallel of $\triangle H^v$ and T_b for nonpolar and slightly polar compounds, as per the *Hildebrand rule*.[2] These data explain qualitatively the significance of both molar volume, V_i, and volume polarizability, α_e^v, to $\triangle H^v$ and T_b.

Electron polarizability, α_e^v, can be related to the refractive index, n, of the compound through the *Lorentz–Lorentz equation* (see Reference 1 for more details):

$$\alpha_e^v = \frac{3}{4}\pi N \frac{n^2 - 1}{n^2 + 2} \quad (4)$$

where N is the Avogadro's number.

As measured by the refractive index [Eq. (4)], α_e^v increases sharply in the sequence *n*-alkanes, cycloalkanes, and aromatics for a constant molar volume.

Table 1 shows the relationship of calculated boiling points, based on a plot of boiling point versus $V_c[(n_c^2 - 1)/(n_c^2 + 2)]^2$ function—where V_c and n_c are molar volume and refractive index, respectively, with the experimental values for a series of polar compounds. The calculated values of T_b are seen to be systematically low, indicating the prominence of intermolecular interactions over and above dispersion forces. As a matter of fact, the difference between experimental and calculated values is a good measure of the comparative importance of these nondispersive interactions, which are discussed next.

B. Nondispersion Forces

The previous discussion suggests that since their dispersion interactions per molecule and boiling points are approximately equivalent, the separation of compounds having like values of V_c and n_c would be difficult to achieve. It is important to remember that there are other attractive interactions that can differ considerably among molecules of comparable size and polarizability. They are often referred to as *specific interactions* because of the discriminating nature of these nondispersive forces, as opposed to the nonspecific or ubiquitous nature of dispersion forces. It is important to develop an understanding of their general nature and potential advantages in different systems.

Specific interactions may be broadly classified as follows:

1. Dipole interactions
2. Hydrogen bonding
3. Various electron donor–acceptor interactions (including ionic or covalent bond formation).

The physical basis of each of these interactions entails varying contributions from well-known electrostatic forces and chemical or covalent bonding. As a result, the corresponding interaction energies may vary from weak physical attractions to strong chemical bonds. Let us review each of these specific interactions.

I. Dipole Interactions

The classical electrostatic theory describes how a molecule with a permanent dipole moment can interact with adjacent molecules: when two neighboring molecules possess a permanent dipole individually, dipole orientation occurs such that the positive head of one dipole is positioned near the negative head of the other dipole. For the greatest energy of attraction, it would be necessary for the two dipole vectors to lie on a straight line, with the positive and negative heads of the dipoles as close as repulsive forces would permit. This optimum arrangement of dipoles, however, is countered by the random thermal movements of individual molecules.

A compromising circumstance arises when, at any given moment, neighboring pairs of dipoles are more regularly arranged for attraction than for repulsion. This leads, on average, to a net attractive energy $(E_{cd})_o$ from dipole orientation (*Keesom forces*; for more details, see Reference 2):

$$(E_{cd})_o = -\frac{2\mu_c^2\mu_d^2}{3kTr^6} \tag{5}$$

where μ_c and μ_d are the permanent dipole moments of neighboring molecules c and d, k is the Boltzmann constant, T is the absolute temperature.

a. Induction and dipole orientation

Various observations that are well supported by theoretical calculations indicate that both induction and dipole orientation have important roles in the behavior of solutions.[3] This is the reason why the energy of interaction of a polar molecule with a surrounding nonpolar phase is usually greater than expected on the basis of dispersion interactions only.

Let us review the interactions of the solutes propionitrile and propane with the nonpolar solvent *n*-hexadecane and the polar solvent hexadecylnitrile in dilute solutions. The dispersion interaction energy of propionitrile should be equivalent in both solvents and comparable to that of propane as a solute, when considered on the basis of molecular size and polarizability. When there are no specific interactions, the K values for the distribution of propionitrile or propane between either of these solvents and the gas phase should be approximately equivalent. Gas chromatographic data at 25° show, however, that the K value for propionitrile in hexadecane is about 10-fold greater than that for propane, and about 100-fold greater in hexadecylnitrile.[4] The greater K value (relative to propane) for propionitrile in hexadecylnitrile is directly related to induction interactions.

b. Debye interactions

A molecule, **c**, having a permanent dipole also induces temporary dipoles in adjoining molecules, **d** (whether or not d possesses a permanent dipole). A net attractive energy $(E_{cd})_i$ is the result of this dipole–induced dipole, or Debye interaction,[2] as expressed by the following:

$$(E_{cd})_i = -\frac{\mu_c^2 \alpha_d}{r^6} \qquad (6)$$

When molecules c and d both have permanent dipoles, the total induction interaction is brought about from Eq. (6) by adding terms for each dipole: $-(\mu_c^2 \alpha_d + \mu_d^2 \alpha_d)/r^6$.

Equations (5) and (6) predict elevated interaction energies and increased boiling points for those compounds with permanent dipole moments (relative to nonpolar compounds), as in the case of pure liquids (**c** = **d**). The effect is likely to be more pronounced as the dipole moment increases. The first group of compounds listed in Table 1, including acetonitrile, demonstrates the importance of these electrostatic interactions as measured by the differences in experimental and calculated boiling points (on the basis of dispersion forces). These boiling-point differences rise in a fairly orderly fashion with molecular dipole moment, as we might anticipate. A perfect correlation should not be expected—for a number of reasons. Equations (5) and (6) in fact relate to distinct bond dipoles within the molecule, in the case of liquids, rather than to the molecular dipole moment in its entirety. For example, an ester group with a dipole moment of 1.7 D is composed of an ether bond (1.2 D) and a carbonyl bond (2.8 D). And the dipoles are situated so that they are partially canceled in the molecular dipole. In contrast, neighboring molecules in liquids tend to interact by way of individual bond dipoles, so that electrostatic interactions in methyl acetate are somewhat larger than that which would be predicted on the basis of its molecular dipole moment.

In molecules with two or more polar substituents, the bond dipole effect can be even more pronounced because the substituent dipoles may be sufficiently far apart to interact independently with adjoining molecules. This is illustrated in Table 1 by another group of compounds—the dichloroethylenes and the isomeric dichlorobenzenes. There is little deviation in boiling point or boiling-point difference in spite of great variances in the molecular dipole moments among each group of isomers. It is not the net dipole moment, but the group dipole moments of these chloro substituents (2.1 D) that control intermolecular interaction. This suggests that the molecular dipole moment of a molecule is not always a good measure of its polarity or tendency to interact with other polar molecules.

2. Hydrogen Bonding

When an atom of hydrogen is attracted by two atoms rather than just one, under certain conditions, it may be considered to be acting as a bond between them. The hydrogen atom is rather unique in that when its electron becomes detached, only the nucleus—having a very small diameter—remains. Hydrogen bonds formed in formic acid or salicylic acid are shown in Figure 3.

In formic or acetic acid, hydrogen bonds formed are sufficiently strong so that an appreciable concentration of the dimerized molecule can be found in the vapor state. In salicylic acid, an intramolecular hydrogen bond is formed (see Fig. 3). This bonding is likely to release the hydrogen of the carboxyl group. This is why salicylic acid is much stronger acid than its meta or para isomers.

For compounds containing —COOH, —OH, and >NH groups, it is frequently necessary to consider hydrogen bonding and its effects. In these cases, a proton-donor group interacts with a proton-acceptor group or an atom with unpaired electrons.

The effect of hydrogen bonding on boiling points can be seen from the last two compounds in Table 1, methyl alcohol and water (Fig. 4). They show quite large deviations from their experimental boiling points to the calculated values, despite rather modest dipole moments (compare these compounds with propyl chloride or ethyl bromide). This behavior is normal for compounds with an OH group and has been ascribed to hydrogen bonding between the molecules as exemplified here.

To form a hydrogen bond, a proton-donor group (e.g., —OH, >NH, —SH, H—CCl$_3$) interacts with a proton acceptor, an atom or group with unpaired electrons (e.g., —O—, =N—, —F, —S—, —Cl,)C=C(, phenyl). The resulting energy of the interaction is usually large when the two atoms attached to the bonding proton are nitrogen or oxygen. Because the basis of hydrogen bonding is largely electrostatic, the normally small O—H or N—H dipole is effectively enhanced by the small radius of the hydrogen atom. This permits a close approach of the hydrogen end of the dipole to an interacting proton acceptor,

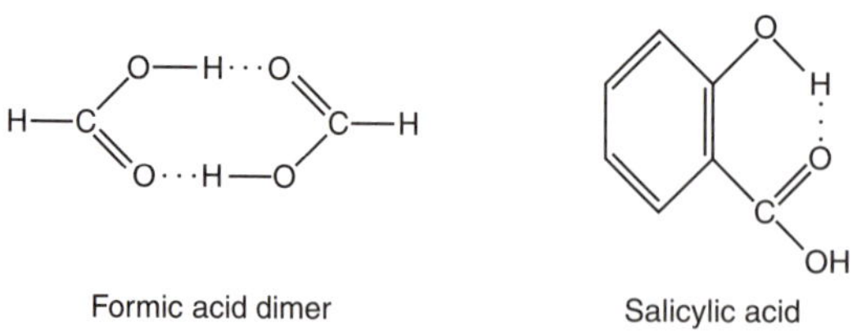

FIGURE 3 Hydrogen bonding in formic and salicylic acids.

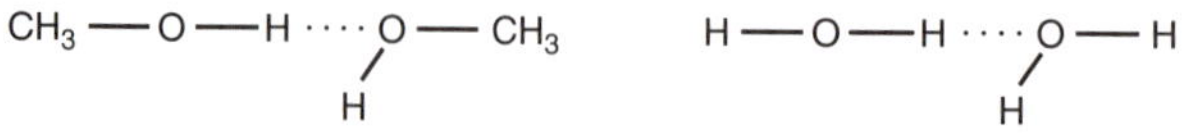

FIGURE 4 Hydrogen bonding in methanol and water.

resulting in a smaller value of r and a greater interaction energy [see Eqs. (5) and (6)].

The energies of hydrogen bonds vary from 4 to 25 kJ/mol. Various factors that determine the energy of a hydrogen bond are not fully understood.[5,6] The bond strength generally grows with increasing basicity of the proton-acceptor group and increasing acidity of the proton-bearing group. The preferred configuration of the hydrogen bond is linear, as shown for the molecules of methanol and water. The basicities of the various acceptor groups are roughly commensurate with the parameter E_B^* of Table 2, declining in this order: amines, neutral oxygen compounds, nitriles, unsaturated hydrocarbons, and sulfides. The acidities of proton donors are similarly proportional to the parameter E_A^* of Table 2, decreasing in the following order: strong acids, $CHCl_3$, phenol, alcohols, and thiophenol (some studies suggest that $CHCl_3$ may be a weaker acid than the alcohols and phenol).

TABLE 2 Characteristics of Various Acids and Bases

Acids, in order of increasing softness	E_A^* (hard)	C_A^* (soft)		Bases, in order of increasing softness	E_B^* (hard)	C_B^* (soft)
HF	17.0	0.0	↑	Ketones, R–CO–R	0.7	0.1
$CHCl_3$	5.1	1.0		Alcohols, R–OH	0.8	0.1
Alcohols, R–OH	3.6	0.8		Ammonia, NH_3	1.3	0.3
Phenol	4.7	5.7		Amides, R–CO–NH_2	1.0	0.3
Thiophenol	1.4	1.7	Hard	Sulfoxides, R–SO–R	1.0	0.3
				Nitriles, R–CN	0.5	0.2
				Esters, R–COO–R	0.6	0.2
				Primary amines, R–NH_2	1.2	0.6
			↓	Ethers, R–O–R	0.7	0.4
SO_2	1.1	7.2	↑	Pyridine	0.9	0.9
			Intermediate	Secondary amines, R2–NH	0.9	0.9
Tetracyanoethylene	1.7	15.0	↓	Benzene	0.1	0.1
Iodine	1.0	10.0	↑	Tertiary amines	0.6	1.2
			Soft	Sulfides	0.0	0.8
			↓	Thiomides	0.1	0.9

Some other acids			Some other bases		
Hard	Intermediate	Soft	Hard	Intermediate	Soft
H^+, Li^+, Na^+, K^+, Be^{2+}, Al^3 Fe^{3+}	Fe^{2+}, Ni^{2+}, Cu^{2+}, Zn^{2+}, Pb^{2+}	Cu^+, Ag^+, Hg^+, Hg^{2+}, Tl^+	H_2O, OH^-, F^-, Cl^-	NO_2^-	Br^-, I^-, SCN^-, olefins

Source: Adapted from References 7 and 8.

3. Various Electron Donor–Acceptor Interactions

It is important to note here that hydrogen bonding is only one example of a class of specific interactions. Various other electron donor–acceptor or acid–base reactions may be represented as follows:

$$A + B \rightleftharpoons A : B$$

Lewis acid–base reactions including charge-transfer complexes, and the combination of ions to form salts or complex ions may also be included in this group.

A comprehensive qualitative theory of the relative strengths of the acid–base bonds in terms of so-called hard and soft interactions has been provided by Pearson.[7] Hard interactions are predominantly considered electrostatic, whereas soft interactions can be regarded as mainly covalent. The acids and bases that can more readily enter into electrostatic interactions are referred to as hard, while soft acids and bases have a greater tendency to form covalent bonds based on orbital overlap. Due to the discrete and fundamentally different nature of these two types of interactions, soft acids prefer to combine with soft bases, and hard acids with hard bases. These acid and base types may be characterized as follows:

- Hard acids have a small size, high positive oxidation state, and no outer electrons that can be excited to a higher state.
- Soft acids have large size, low or zero positive charge, and several easily excited outer electrons.
- Hard bases have low polarizability, high electronegativity and are hard to reduce; empty orbitals have high energy and hence are inaccessible.
- Soft bases have high polarizability and low electronegativity, are easily oxidized, and have empty low-lying orbitals.

Table 2 provides several examples of hard and soft acids and bases. In general, softness increases regularly as we move down the periodic table within a particular group (e.g., among halogens F is hard while I$^-$ is soft).

a. Molecular complexes

Hydrogen bonding may be considered a relatively hard acid–base interaction, and strong hydrogen bonding usually requires a hard proton donor and acceptor. By contrast, molecular complexation of the picric acid–aromatic hydrocarbon type is a soft interaction, in which compounds such as picric acid, trinitrofluorenone, and tetracyanoethylene play the role of soft acids, and several aromatic compounds act as soft bases. The interactions of silver ion or mercuric ion with organic compounds may be considered soft; this helps explain why the Ag$^+$-olefin, Ag$^+$-aromatic, and Hg^{2+}-alkylsulfide complexes are relatively stable.

b. Ion exchange

The selectivity of ion-exchange processes seems to be controlled partly by the criterion of hard versus soft interactions. In cation exchange, the distribution of ions A$^+$ and B$^+$ between a sulfonic acid resin and the surrounding water phase is governed by the relative stability of resin and water complexes of the two ions:

$$\text{Resin–SO}_3^- A^+ + B^+(H_2O) \rightleftharpoons \text{Resin–SO}_3^- B^+ + A^+(H_2O)$$

Only a single, exchangeable c is shown here; however, many water molecules presumably interact with cations in either phase.

Here the resin–SO$_3^-$ groups may be considered as bases that compete with water of hydration in the encompassing aqueous phase for the cationic acids A$^+$ and B$^+$. Because water is one of the hardest bases, we expect it to interact preferentially with the hardest acids or cations. The selectivity or the preferential adsorption of a cation on the resin rises with increasing softness of the cation: Li$^+$ < H$^+$ < Na$^+$ < K$^+$ < Rb$^+$ < Cs$^+$ < Ag$^+$ < M$^-$ (see Table 2, and recall that softness increases as we go down the periodic table). This is also the case with divalent cations, Mg^{2+} < Ca^{2+} < Sr^{2+} < Ba^{2+} < Pb^{2+}. These trends in the stability of ion–ion-exchange resin complexes may be clarified in equivalent terms by means of classical electrostatic theory and the charge/radius ratio of competing ions.[8]

c. Interaction energy of various acid–base interactions

The interaction energy E_{ab} of these acid–base interactions can be expressed roughly[6] as follows:

$$E_{ab} = E_A^* E_B^* + C_A^* C_B^* \qquad (7)$$

Here E_A^* and E_B^* measure the electrostatic acid and base strength of a given molecule, respectively, and C_A^* and C_B^* are corresponding measures of covalent acid and base strengths. Table 2 lists the values of these parameters for several acids and bases. The separation of a soft from a hard base in an equilibrium distribution is seen to be preferred if one of the two phases contains a very hard or soft acid that will interact selectively with one of the two compounds to be separated, that is, in the case of secondary chemical equilibria.

The total energy of interaction of a molecule c with a surrounding phase d is the total of all possible interactions: dispersion E_d, dipole orientation E_o, and induction E_i (for all adjacent molecules d), and acid–base interactions. Remembering that the individual interactions are added over all atoms or groups in the molecule, we may write this total interaction energy E_{cd} as follows:

$$E_{cd} = E_d + E_o + E_i + E_{ab} \qquad (8)$$

In situations where we have large multifunctional molecules that are often encountered in real separation problems, Eq. (8) becomes rather complex. Also, we usually do not know the numerical values of the various parameters in Eq. (8). Moreover, the equation is at best approximate, and accurate estimation of intermolecular interaction energies are often impossible. The principal benefit of this equation and the relationships that it summarizes is for qualitative predictions and also for correlating and generalizing experimental findings. Additionally, Eq. (8) establishes the most important

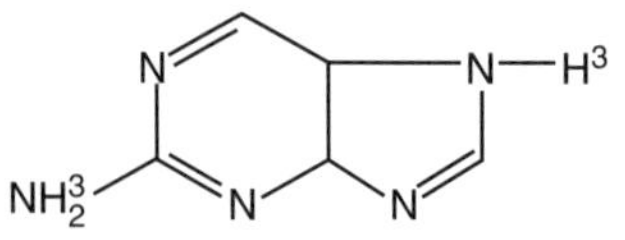

FIGURE 5 Structure of the labeled compound.

factors that govern interaction energy and separation in phase equilibrium and distribution systems: dispersion interactions, electrostatic interactions involving permanent dipoles, and the electron donor–acceptor properties of different molecules [Eq. (7)]. When two compounds are considerably unlike in any of these three properties, their separation is generally quite direct. The preceding discussion provides broad guidelines for the choice of the best phases in an equilibrium separation process in such cases.

It may not be obvious in some cases that two closely related compounds are considerably different in polarity, molecular size, or electron-sharing properties. However, it is an exceptional situation when two such compounds prove inseparable in every equilibrium distribution system. The reason is that essentially every molecular difference inevitably has an effect on the phase-distribution properties of the molecule, and modern separation methods (notably chromatography, as discussed in Chapter 6) are capable of taking advantage of remarkably small differences in molecular distribution to achieve total separation of species. It is even possible to separate large molecules that differ only in the sign of an optically active center or in an isotopic label. For example, the tritium-labeled compound shown in Figure 5 can be separated from its unlabeled derivative.[9]

The preceding discussion emphasizes physical interactions, as opposed to strong chemical bonding. Strong chemical interactions are commonly useful in a separation system (e.g., reaction of acids with bases). However, it is important to remember that a chemical reaction is often effectively irreversible under the conditions of separation, and irreversibility or slow reversibility can be highly detrimental in an equilibrium distribution system because it severely restricts separation effectiveness in multistage operations.

II. SOLUBILITY PARAMETER THEORY

An ideal solution is generally used as a model for understanding and predicting the behavior of a system of interest. A very important assumption is made for an ideal solution that all intermolecular interactions are equivalent, and thus the heat of mixing is zero. However, as can be seen by the preceding discussion, this assumption may frequently be unreliable. As a matter of fact, unequal interactions between molecules appear to account for most of the apparent differences in equilibrium distribution systems. For regular solutions, the theory focuses on these unequal intermolecular forces and recognizes that the heat of mixing is not necessarily zero.[2, 10] In the theory of regular solutions, we assume that the entropy of mixing is ideal and, in its simplest form, no volume change occurs on mixing. It is further assumed that only dispersion forces are significant. As can be seen from previous discussion, this is a major approximation and as a result, subject to errors.

The chemical potential of a component c of the regular solution is obtained from the equation for an ideal solution ($\mu_c - \mu_c^0 = RT \ln X_c$; where, X_c is the mole fraction of solute c in the final solution) by adding the enthalpy term ($h_c - h_c^0$):

$$\mu_c - \mu_c^0 = (h_c - h_c) + RT \ln X_c \tag{9}$$

Using chemical potential ($\mu = \mu_0 + RT \ln a$) and activity a (equal to γX_i), we obtain the following equation:

$$\mu_c - \mu_c^0 = RT \ln \gamma X_c \tag{10}$$

which, when combined with Eq. (9) gives

$$RT \ln \gamma = h_c - h_c^0 - \Delta h^e \tag{11}$$

where $\triangle h^e$ is the partial molar *excess* enthalpy

Let us relate the enthalpy term ($h_c - h_c^0$) to particular physical properties of the system. We can start by considering the energy of interaction of a molecule c in pure c versus that of c in a dilute solution of some solvent j. This will lead us straight to the heat of combining c with j to form a dilute solution. The following derivation is a simplified version of that often used in the development of Hildebrand's *solubility parameter theory*, one of the major practical uses of regular solution theory (see Reference 11 for more details). However, the final expression we obtain will be the same as that provided by the original, more precise approach.[2,10] It is anticipated that this derivation will promote appreciation of the physical justification and limitation of solubility parameter theory.

The solubility parameter theory as advanced initially assumes that only dispersion interactions in systems have to be considered, since they are the important factors. This provides a good approximation for solutions of nonpolar or weakly polar compounds, but is less valid for other systems. For the purpose of the next discussion, we are going to assume that only dispersion interactions are occurring.

Let us look at a molecule c in the pure liquid phase, surrounded by other molecules c, and assume some number (e) of dispersive interactions between c and adjacent molecules. For clarity, we will also suppose that only interactions between nearest neighbors are important, and that c has e nearest neighbors:

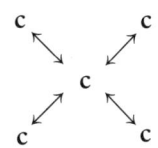

$$\begin{matrix} c & & c \\ \searrow & & \swarrow \\ & c & \\ \swarrow & & \nwarrow \\ c & & c \end{matrix}$$

Furthermore, we will assume that each interaction is of constant energy E_{cc}. The energy of vaporization of a molecule c may be envisioned as the outcome of two sequential processes: elimination of c from the liquid phase,

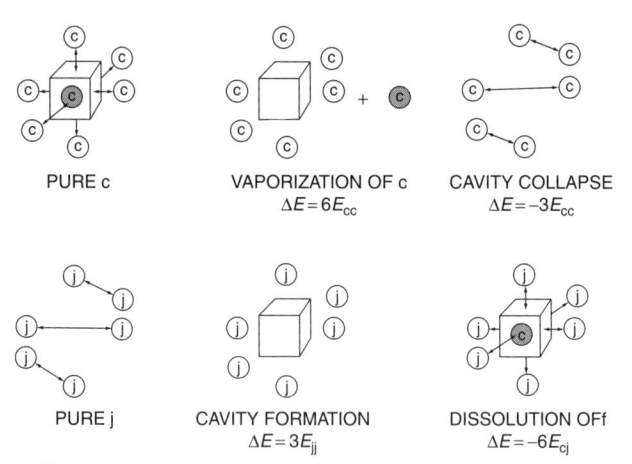

FIGURE 6 Visualization of mixing of c with j (e = 6) in a regular solution.

succeeded by disintegration of the cavity initially occupied by c. This is illustrated in Figure 6 for $e = 6$.

The energy of elimination (eE_{cc}) is the amount of energy needed for breaking e c–c interactions between c and neighboring molecules. The disintegration of the resulting cavity demands that each of the molecules initially adjoining c forms a new c–c interaction, producing $e/2$ for interactions of this type. The subsequent total energy (cavity disintegration) is $-(e/2)E_{cc}$. The net energy of vaporization ΔE_c^v is the total amount of the energies for these two processes times the number of molecules per mole (N).

$$\Delta E_c^v = N(e/2)E_{cc} \tag{12}$$

Let us now think about the dissolution of the vaporized molecule c in pure j. This process may be perceived as the opposite of vaporization–cavity production, followed by insertion of c into the cavity. We may now presume an equal number of intermolecular interactions per unit volume, so that $e/2$ j–j interactions will be broken in creation of the cavity (a number equal to that formed in the disintegration of the previous cavity of equal size). In a like manner, e c–j interactions will be forged on insertion of c into the cavity. If the energies of these particular interactions are labeled E_{jj} and E_{cj}, respectively, the total energy of dissolution of vaporized c and j (per mole of c) will be

$$\Delta E_{cj}^S = N\left[\frac{e}{2}E_{jj} - eE_{cj}\right] \tag{13}$$

Assuming an equal number of intermolecular interactions per unit volume, Ne is proportional to V_c: $Ne = CV_c$, where C is a proportionality constant and V_c is the molar volume of c. Insertion of this expression into Eq. (12) gives

$$\Delta E_c^v = \left[\frac{C}{2}\right]V_c E_{cc} \tag{14}$$

We can now define a meaningful molecular property, the solubility parameter δ_c of compound **c**:

$$\delta_c^2 = \frac{\Delta E_c^v}{V_c}$$

$$= \left[\frac{C}{2}\right] E_{cc} \qquad (15)$$

where δ^2 gauges the intermolecular interaction energy per unit volume in the pure liquid and is proportional to E_{cc}. Similarly, E_{cc} is proportional to E_d (in dispersion interactions only), so that from Eq. (15), we see that for nonpolar compounds δ_c should be proportional to the polarizability per unit volume of $c(a_e^v)$. We can also express E_{cc} in terms of the solubility parameter for **j**: $\delta_j^2 = (C/2)E_{jj}$. Another assumption of the solubility parameter theory is that E_{cj} is the geometric mean of E_{cc} and E_{jj}: $E_{cj} = \sqrt{E_{cc}E_{jj}} = (2/C)\delta_c\delta_j$.

The total energy of combining a mole of **c** with a large quantity of pure **j** (to form a dilute solution of **c** and **j**) is now shown as the sum of ΔE_c^v [Eq. (12)] and ΔE_{cj}^s [Eq. (13)]:

$$\Delta E_{cj}^M = N\left[\frac{e}{2}\right](E_{cc} - 2E_{cj} + E_{jj})$$

$$= V_c(\delta_c^2 - 2\delta_c\delta_j + \delta_j^2) \qquad (16)$$

Recalling that $\Delta V^M = 0$, so that $\Delta E_{cj}^M = \Delta H_{cj}^M = h_c - h_c^0$ we have

$$h_c - h_c^0 = V_c(\delta_c - \delta_j)^2 \qquad (17)$$

Here $h_c - h_c^0$ is the heat of combining a mole of pure **c** with a large amount of **j** to form a dilute solution. Equation (17) is of essential importance in its capability to quantitatively account for a substantial scope of distribution phenomena and also for the fundamental understanding it affords in the combination of substances that are not similar. The compound **c** of Eq. (16) can be a solid, liquid, or gas.

When the concentrations of **c** and **j** in solution are both significant (i.e., it is not a dilute solution), a similar relationship exists as demonstrated in the following equation:

$$\Delta H_{cj}^M = (X_c V_c + X_j V_j)(\delta_c - \delta_j)^2 \Phi_c' \Phi_j' \qquad (18)$$

where ΔH_{cj}^M is the heat of mixing per mole of final solution; X_c and X_j are the mole fractions of **c** and **j** in the final solution; V_c and V_j are the molar volumes of **c** and **j**; Φ_c' and Φ_j' are the volume fractions of **c** and **j**.

When the mole fraction of **c** becomes small, Eq. (18) is reduced to Eq. (17).

The deviations from ideality in the combining of two substances, **c** and **j**, are related to the intermolecular forces in the pure liquids according to Eq. (17), as measured by the solubility parameter, δ. As a result, the solubility parameter of a specific liquid is a significant index to its performance in distribution processes. The values of δ for some common solvents are shown in Table 3 (see also Reference 12). Large δ values reflect large interaction energies in the pure liquids. These arise from strong dispersion interactions in some cases

TABLE 3 Solubility Parameters of Some Common Solvents (at 25 °C)

Compound	δ	Compound	δ
n-Pentane	7.1	Acetic acid	12.4
Cyclohexane	8.2	Methanol	12.9
Benzene	9.1	Formamide	17.9
		Water	21.0

(e.g., CS_2, CH_2I_2), but usually a large δ value indicates the existence of effective specific interactions. It is useful in these cases to subdivide the solubility parameter as a whole into contributions from dispersion, dipole, and hydrogen-bonding interactions.

The heat of mixing is always ≥ 0, according to the solubility parameter theory. As a result, $\triangle H^M$ is always unfavorable for mixing, becoming more so as the two liquids become more unlike in terms of their δ values. This is the physical foundation of the well-known rule, "like dissolves like." When the δ values of two liquids become adequately dissimilar, mixing is largely precluded and two phases can coexist.

The difference in chemical potentials of a compound c in a pure liquid (μ_c^0) and as a dilute, regular solution in j (μ_c) is obtained from Eqs. (9) and (16):

$$\mu_c - \mu_c^0 = V_c(\delta_c - \delta_j)^2 + RT \ln X_c \qquad (19)$$

Admixture is favored by the entropy term, $RT \ln X_c$, and is countered by the heat term, $V_c(\delta_c - \delta_j)^2$. The solubility of a liquid c in another liquid j can be determined from Eq. (19) because at equilibrium $\mu_c - \mu_c^0$ is zero. But the heat term must be replaced by that given in Eq. (17) when the solubility of c in j is substantial. The activity coefficient γ_c for c dissolved in j may be derived from Eqs. (11) and (16):

$$\ln \gamma_c = \frac{V_c(\delta_c - \delta_j)^2}{RT} \qquad (20)$$

In ideal solutions, from the definition of an activity coefficient, $\gamma_c = 1$. Thus the definition of an ideal solution in terms of solubility parameter theory is $\delta_c = \delta_j$. Careful scrutiny of Eq. (20) verifies that all deviations from Raoult's law in regular solutions will be positive.

The activity coefficient of a solute is very meaningful in accounting for distribution between two phases. Thus, it is useful to know how γ varies with the difference in solute and solvent solubility parameters and how it varies with solute molar volume. The solubility of a solid c in a nonideal solvent j may be derived from the instance of an ideal solution. Solubility changes reciprocally with the activity coefficient, so that

$$\ln X_c = \frac{\Delta H^F(1/T_f - 1/T)}{\gamma R} \qquad (21)$$

The supposition that only dispersion forces are included in the various intermolecular interactions was the basis of the original derivation of the solubility parameter theory. For compounds that also interact by definite forces, Eq. (16) is ordinarily less accurate. The failure of the geometric mean rule (for dispersion forces only) is anticipated in mixtures of polar and nonpolar molecules, since dipole orientation and acid–base type interactions are precluded between solute and solvent. Inadequacy of the geometric mean rule is especially conspicuous in the interaction of hydrogen-bonding solutes and solvents and other acid–base pairs. Negative deviations from Raoult's law and activity coefficients of less than 1 are often the result when such compound types combine with the evolution of heat. An expanded version that recognizes different types of interactions can partially correct for the breakdown of the classical solubility parameter treatment.

III. GROUP INTERACTIONS

Our discussion of intermolecular interactions in the previous section indicated that the total interaction energy E_{cj} of a molecule, c, with a surrounding liquid phase, j, is the total amount of the interactions for each group x, y, z, etc. present in the molecule c.[1] For instance, let us review the case of a dilute solution of propionaldehyde (c) in n-hexadecane (j). For a propionaldehyde molecule CH_3CH_2CHO, E_{cj} is the sum of interaction energies for a methyl group ($E_{CH_3\text{-}j}$), a methylene group ($E_{CH_2\text{-}j}$), and a nitrile group ($E_{CHO\text{-}j}$) with surrounding solvent j.

$$E_{cj} = E_{CH_3\text{-}j} + E_{CH_2\text{-}j} + E_{CHO\text{-}j}$$

This equation may be applied to any molecule c and any phase j as follows:

$$E_{cj} = \sum^{x} E_{x\text{-}j} \qquad (22)$$

where $E_{x\text{-}j}$ is the total interaction energy of a group x in the molecule c and the surrounding phase is j. The total amount is taken over all groups x in the molecule c.

The difference in interaction energies of c with each phase ($E_{ck}-E_{cj}$), for the distribution of c between two phases j and k, will be roughly the difference in partial molar enthalpies ($h_{ck}-h_{cj}$), and a comparable equality will exist between the group interaction energies, $E_{x\text{-}j}$ and $E_{x\text{-}k}$, and the corresponding enthalpy terms, $h_{x\text{-}j}$ and $h_{x\text{-}k}$. As a result, we can write

$$h_{c\text{-}k} - h_{c\text{-}j} = \sum^{x}(h_{x\text{-}k} - h_{x\text{-}j}) \qquad (23)$$

Recognizing that $K_x = \gamma^y/\gamma^x$ and relating to Eq. (11), we have

$$\ln K_x = \frac{\sum^{x}(h_{x\text{-}k} - h_{x\text{-}j})}{RT}$$

$$\log K_x = \sum^{x} f(x = jk) \qquad (24)$$

The group parameter $f(\mathbf{x}-\mathbf{j}, \mathbf{k})$ is equal to $(h_{x-k}-h_{x-j})/2.3RT$ and is, as a result, a function of the group $\mathbf{x}$ and the compositions of phases $\mathbf{j}$ and $\mathbf{k}$. Based on Eq. (24), the logarithm of K_x (or K) for distribution of a molecule $\mathbf{c}$ between two phases can be described by the sum of the group constants $f(\mathbf{x}-\mathbf{j}, \mathbf{k})$ for all of the groups in the molecule.

In the given set of phases $\mathbf{j}$ and $\mathbf{k}$ in a separation system of interest, $f(\mathbf{x}-\mathbf{j}, \mathbf{k})$ is assumed to have the same value for a given group $\mathbf{x}$, whether $\mathbf{x}$ is a substituent in different molecules or several groups $\mathbf{x}$ are present in the same molecule. This indicates that the addition of a particular group to any molecule should result in a constant increase in $\log K$ as exemplified by addition of $-NH_2$ to benzene to yield aniline or addition of another $-NH_2$ to aniline to produce diaminobenzene. Since a large number of different molecules can be produced from a relatively small number of different groups $\mathbf{x}$, Eq. (24) is valuable for correlating and predicting K values in distribution systems as a function of the molecular structure of the solute $\mathbf{c}$.

A variation of Eq. (24) was first derived by Martin[13] for the correlation of distribution constants in his paper on chromatography; and it is frequently referred to as the *Martin equation*.

The reliability of the Martin equation in predicting K values is based on the constancy of $f(\mathbf{x}-\mathbf{j}, \mathbf{k})$ values for a given group $\mathbf{x}$ in different molecules or for different positions in the same molecule. However, it is important to remember that this constancy of the group parameters requires strict comparability of equivalent groups $\mathbf{x}$ in regard to their interaction with a surrounding phase. The structural equivalency is not always sufficient in this respect, as illustrated by the two compounds phenol and *o*-nitrophenol. It might be assumed that the interaction energies of the $-OH$ group with the surrounding phase are the same in both compounds. However, this interaction energy varies with dipole moment and acidity of the $-OH$ group. Because of intramolecular hydrogen bonding and resonance interactions between the $-OH$ and $-NO_2$ groups in *o*-nitrophenol, the dipole moment and acidity of this $-OH$ group are significantly different from the values in phenol. As a result, $f(\mathbf{x}-\mathbf{j}, \mathbf{k})$ will not be the same for the $-OH$ groups in these two compounds, thus suggesting an apparent failure of the Martin equation. Therefore, the Martin equation should be mainly applied to compounds in which the various groups $\mathbf{x}$ are carefully defined, taking into account any adjacent groups that are likely to affect each other's interactions with the surrounding phase. Equation (24) is quite reliable for a homologous series $CH_3-(CH_2)_n-Y$, since for $n \leq 1$ each additional methylene group is essentially equivalent in regard to its interactions with the surrounding phase. It can yield a general relationship for a homologous series R—Y:

$$\log K = f(CH_3-j, k) + f(Y-j, k) + nf(CH_2-j, k)$$
$$= A + Bn \qquad (25)$$

A and B are constants for a given homologous series and a particular separation system; therefore, $\log K$ increases linearly with carbon number $(n+1)$.

Equation (24) can be also used to include structural variations in the molecules that constitute the surrounding phase (see Reference 14). Various interactions of phase **j** with a particular solute group **x** can be broken down into interactions of **x** with different groups in a molecule **j**, in the same way as different groups in **c** are handled. This approach is very complicated. As a result, it is less widely used than the Martin equation.

REFERENCES

1. Karger, B. L., Snyder, L. R., and Horvath, C. *An Introduction to Separation Science*. Wiley, New York, NY, 1973.
2. Hildebrand, J. H. and Scott, R. I. *The Solubility of Non-electrolytes*. Dover Publications, New York, NY, 1964.
3. Meyer, E. and Ross, R. A. *J. Phys. Chem.* 75:831, 1971.
4. Kovats, E. and Weicsz, P. B. *Ber. Bunsengesell.* 69:812, 1965.
5. Taft, R. W., Gurka, D., Joris, L., von, P., Schlyer, R. and Rakshys, J. W. *J. Am. Chem. Soc.* 91:480, 1969.
6. Drago, R. S., Vogel, G. C., and Needham, T. E. *J. Am. Chem. Soc.* 93:6014, 1971; also see 86:5240, 1964.
7. Pearson, R. G. *Science* 151:172, 1966.
8. Helfferich, F. *Ion Exchange*, p. 158. McGraw-Hill, New York, NY, 1962.
9. Klein, P. D. *Advances in Chromatography* (J. C. Giddings and R. A. Keller, Eds.), Vol III. Marcel Dekker, New York, NY, 1966.
10. Hildebrand, J. H. and Scott, R. L. *Regular Solutions*. Prentice-Hall, Englewood Cliffs, NJ, 1962.
11. Pitzer, K. S. and Brewer, L. *Thermodyanamics*. McGraw-Hill, New York, 1961.
12. Hoy, K. L. *J. Paint Tech.* 42:76, 1970.
13. Martin, A. J. P. *Biochem Soc. Symp.* 3:4, 1949.
14. Redlich, O., Derr, E. L. and Pierotti, G. J. *J. Am. Chem. Soc.* 81:2283, 1959.

QUESTIONS FOR REVIEW

1. Define and explain the following:
   a. Intermolecular interactions
   b. Debye interactions
   c. Dispersion interactions
   d. Specific interactions
2. Describe briefly the solubility parameter theory for regular solutions.
3. Derive the Martin equation, and explain its importance.

5
MASS TRANSPORT AND SEPARATION

I. TYPES OF DIFFUSION
II. MODELING DIFFUSION
   A. Fick's First Law
   B. Fick's Second Law
III. CHROMATOGRAPHIC IMPLICATIONS
   A. Longitudinal Molecular Diffusion
   B. Sorption and Desorption
   C. Flow and Diffusion in the Mobile Phase
IV. DIFFUSION RATES IN VARIOUS MEDIA
   A. Diffusion in Gases
   B. Diffusion in Liquids
   C. Diffusion in Solids
V. MASS TRANSFER
   A. Flow Through Packed Columns
   REFERENCES
   QUESTIONS FOR REVIEW

To separate a mixture successfully, it is necessary to transport mass from one location to another. It is well known that in this process, diffusion plays a significant role because dissolved substances diffuse from a region of higher concentration to a region of lower concentration. The diffusion process would continue, if left undisturbed, until uniform concentration is achieved throughout the system.

Diffusion is closely related to Brownian motion that leads to the consideration that the molecules or particles of a substance diffuse because of their Brownian motion (named after Robert Brown, an English botanist). In 1827 he observed under a microscope that very small particles suspended in a liquid were in a state of ceaseless "erratic" motion. He interpreted the results of this random motion as follows: "the movements arose neither from currents in the fluid nor from its gradual evaporation, but belonged to the particle itself."

Brownian motion and diffusion can be defined as follows:

Brownian motion: Brownian motion is the motion resulting from the sum total of the impacts of molecules on the sides of small particles that do not cancel out at every instant.

Diffusion: Diffusion can be defined as a mass transport on a molecular scale.

It is important to note that in diffusion we are interested in motion at the molecular level.

Diffusion is sometimes called bulk motion or convection, depending on the scale and type of instruments that are being used.[1] As mentioned above, separation requires that mass be transported from one location to another. It is the diffusion process that makes this possible. Furthermore, this process is spontaneous and is accompanied by a decrease in free energy.

Mass transportation on an intermediate scale may also occur via turbulent eddies; therefore, their effect also has to be considered. In general, separation is favored under those conditions, where molecules of one component transfer to a greater extent than those for another. These transfers take place by diffusion because other transport processes do not differentiate between molecules of different kinds. This explains why diffusion is the most important transfer mechanism for achieving a separation. It is important to remember that bulk flow and mass transfer due to convection may create the essential concentration gradient for diffusion and thus play a significant role in the separation process.

In Chapter 4, we discussed equilibrium aspects of separation without considering the speed of achieving equilibrium in a particular phase. Here we will focus on the dynamic aspects that influence the speed, and to a great extent the efficiency, of the equilibrium separation process. It should also be noted that many separation techniques are of a solely dynamic nature; that is, the separation is achieved by different transport rates of the components. As a result, any discussion of these rate processes, for example, thermal diffusion or electrophoresis, should be based on the rudimentary mass transport phenomena.

I. TYPES OF DIFFUSION

As mentioned earlier, mass transport by individual molecular motion is generally called diffusion. To achieve useful separations, the diffusion of a particular species in a mixture, that is, interdiffusion is of greater interest as opposed to self-diffusion in a system having identical molecules only.

Let us review different types of diffusion.

- *Molecular Diffusion*. A commonly encountered diffusion process where the concentration gradient, namely, the chemical potential gradient, is the driving force of molecular motion. It is called ordinary, or more appropriately, molecular diffusion.
- *Pressure Diffusion*. In some separation processes, the molecular motion of the species may also be caused by a pressure gradient, bringing about pressure diffusion.
- *Thermal Diffusion*. When the molecular motion of the species is caused by a thermal gradient, the diffusion process is called thermal diffusion.
- *Forced Diffusion*. The diffusion resulting from an external field acting on the component of interest is called forced diffusion.

Electromigration is an example of forced diffusion because a force field is applied to achieve the desired result. Forced diffusion is exploited in electrophoresis to achieve very useful separations. The diffusion in a gravitational field is another example of forced diffusion that has been found tremendously useful in centrifugation.

- *Diffusion in Chromatography.* It is important to remember that the term diffusion is used mainly for the real molecular processes. In chromatographic systems that are based on the flow of the mobile phase, complex diffusional and flow processes are encountered that are more appropriately called *axial dispersion.*
- *Axial Dispersion.* When mass is distributed in the direction of flow because of complex diffusion and flow processes, the phenomenon is called axial dispersion.

II. MODELING DIFFUSION

The diffusion process can be simply described as a one-dimensional model in terms of Fick's first or second law.

A. Fick's First Law

The diffusional movement of mass in one direction through a plane is given by Fick's first law. In 1855 Fick postulated that for diffusion in one direction, the flow, J, of a substance through a plane perpendicular to the direction of diffusion is directly proportional to the rate at which concentration changes with distance, dC/dx, the concentration gradient, as shown in the following equation:

$$J = -D\frac{dC}{dx} \qquad (1)$$

where $J =$ flux (the rate of mass flow per unit area in the direction x in $g\,mol/cm^2/s$), $D =$ diffusion coefficient in cm^2/s, $C =$ concentration of the solute in $g\,mol/cm.^3$

In Eq. (1), dc/dx is the concentration gradient ($g\,mol/cm^4$) in the direction x, which is normal to the plane. The mass flow in the direction of lower concentration is indicated by the negative sign. As per this equation, the flux is directly proportional to the concentration gradient as it is equal to the product of driving force (concentration gradient) and the diffusivity constant.

The diffusion coefficient is not always constant, as it can depend on concentration. Therefore, the actual driving force is the gradient of chemical potential, which may be different from the concentration gradient. However, in dilute solutions, Eq. (1) is valid. It should be noted that in a single phase, diffusion takes place in any direction in which there is a concentration gradient. The three-dimensional nature of molecular diffusion is undesirable from the standpoint of separation technology because it leads to mixing instead of separation.

B. Fick's Second Law

For solving most diffusion problems, Fick's first law is insufficient because the concentrations are usually unknown. The mathematical treatment of diffusion processes in these cases is better handled by the equation known as Fick's second law:

$$\frac{dC}{dt} = D\frac{d^2C}{dx^2} \qquad (2)$$

where C = concentration and t = time.

According to this equation, the time rate of change of concentration is proportional to the spatial rate of change in the direction of concentration gradient. When both flow and diffusion occur, the time rate of concentration change is given for the one-dimensional case by the following equation:

$$\frac{dC}{dt} = -v\frac{dC}{dx} = D\frac{d^2C}{dx^2} \qquad (3)$$

where v is the flow velocity in x direction.

It is important to note that the information provided by the above equations is limited by the general nature of partial differential equations. A more reliable solution to a given problem may be obtained by using specific information available for that problem and appropriate mathematical methods. The interested reader may want to review the original work of Crank[2] on solution of diffusion equations for a large number of problems as well as very interesting discussion by Giddings.[3]

III. CHROMATOGRAPHIC IMPLICATIONS

It is important to recall from the previous discussion that dissolved substances diffuse from a region of higher concentration to a region of lower concentration. As a result, diffusion is a spontaneous process that is accompanied by a decline in free energy. As mentioned before, diffusion is closely related to Brownian motion that leads to the consideration that the molecules or particles of a substance diffuse because of their Brownian motion. Furthermore, the tendency of a substance to diffuse has been described by Fick's laws. The approximate values of diffusion coefficients of gas, liquid, and supercritical fluid are given below to indicate the impact of diffusion on mass transportation in these modes of chromatography:

Gas	10^{-1} cm^2/s
Liquid	10^{-5} cm^2/s
Supercritical fluid	10^{-3}–10^{-4} cm^2/s

In chromatography, two contrasting approaches, namely, the macroscopic viewpoint, which describes the bulk concentration behavior, versus the microscopic viewpoint dealing with molecular statistics, have to be considered.[3] Each molecule in a chromatographic process is moving along in an erratic jerky motion. The molecule's irregular motion is determined by several independent statistical processes that control zone spreading. Some of these relevant processes are as follows:

- Longitudinal molecular diffusion
- Sorption and desorption
- Flow and diffusion in the mobile phase

A. Longitudinal Molecular Diffusion

Solute molecules are involved in endless Brownian motion, which is responsible for diffusion. The component of this erratic motion along the column axis, superimposed on the downstream displacement caused by flow, is one of the sources of band broadening.

B. Sorption and Desorption

Every time a molecule is sorbed by the stationary phase, its downstream motion ceases. When it is desorbed, the molecule proceeds again. The process of sorption and desorption occurs randomly, thus making this stop-and-go sequence an extremely erratic process.

C. Flow and Diffusion in the Mobile Phase

The microscopic flow process in a packed bed is very tortuous, with each stream path frequently changing direction and velocity. A solute molecule conveyed in this flow travels a very uneven path. The randomness of the molecular path is amplified by Brownian displacements of the molecule from one stream to another.

A one-dimensional diffusion equation for the following two examples will help demonstrate the usefulness of this equation for simple systems:

- Diffusion of matter in one direction from a planar source
- Diffusion from a continuous (extended) source

The results form the basis of theoretical treatment of a band or front dispersion in elution and frontal chromatography, where the effect of flow must be also considered. (In this technique, the solute mixture is introduced as a continuous stream, and a partial separation results from different retardation rates of various components. A more detailed explanation is discussed in Chapter 6.)

Diffusion from one planar source is given by the following equation (for derivation of the equation, see Reference 1):

$$C = \frac{1}{\sigma(2\pi)^{1/2}} e^{-1/2(x/\sigma)2} \qquad (4)$$

where C = concentration in mass per unit length at position x after time t and σ = standard deviation (in units of length) in normalized form.

In chromatography, the spreading of a band or front (see Fig. 1) that moves down the column can be treated as a one-dimensional process. If the

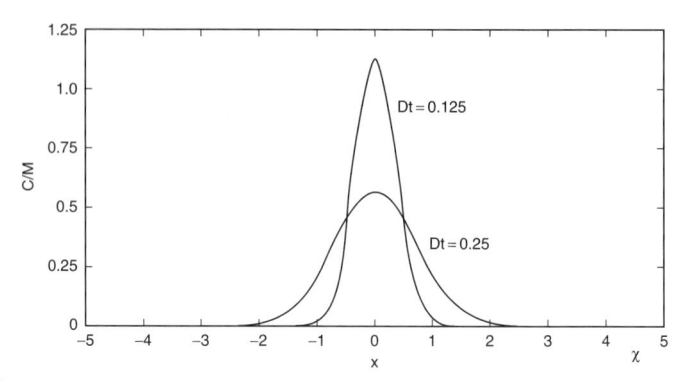

FIGURE 1 Concentration profiles obtained by diffusion from planar source (1).

center of a band or front moves with a velocity, v, we can consider the spreading to take place in a moving coordinate system by the following equation:

$$x' = x + vt \qquad (5)$$

where $x =$ diffusion of solute and $x' =$ position at time t.

Based on Eq. (4), we can derive

$$C = \frac{1}{\sigma(2\pi)^{1/2}} e^{-1/2[(x'-vt)/\sigma]^2} \qquad (6)$$

This equation conveys the fact that the center of distribution is not at $x' = 0$, but is at $x' = vt$ after time t at a flow velocity v.

It is important to note that this equation describes the Gaussian concentration peaks that are usually obtained in elution chromatography. The solute is injected as a planar source into the stream; it progresses down the column and then becomes dispersed in a diffusion-like course. The concentration of the emerging analyte is measured at the exit of the column, and the peak is recorded. The ordinate denotes concentration, C, expressed as amount per volume, and the abscissa represents the time, t, on the recorder chart paper. The Gaussian peak on the chromatogram is basically the graphical expression of Eq. (6).

To evaluate a given peak, we have to consider that x' in Eq. (6) is equal to the length of the column and the concentration is given in mass per unit length. From the recorded chromatogram, we can measure the time, t_R, in which the source moving at the peak center passes through the column at constant velocity, v. Thus, $x' - vt$ may be replaced by $t_R - t$ in the exponent, and the standard deviation may be measured in time units, σ_t. The concentration would then be obtained per unit time.

IV. DIFFUSION RATES IN VARIOUS MEDIA

Discussed below is diffusion in various media such as gases, liquids, and solids. Of these media, diffusion in gases and liquids is of greater interest to the separation scientists.

A. Diffusion in Gases

Diffusion in a gas phase is of interest in separation methods such as distillation or gas chromatography. Interdiffusion of equal molecules (self-diffusion) and interdiffusion of similar small molecules such as nitrogen gas are well understood on the basis of the kinetic theory of gases. However, gas mixtures that contain large molecules and exhibit nonideal behavior are often important. Based on experimentally measured diffusivities, the magnitude of diffusion coefficients are shown in Table 1.

It is generally difficult to obtain diffusion data experimentally. As a result, a dependable estimation of diffusion coefficients in mixtures and their reliance on temperature and pressure is important for quantitative treatment of many separation processes.

Based on kinetic theory, Hirschfelder and coworkers derived the following equation for diffusion constant D_{AB} for a binary mixture:

$$D_{AB} = \frac{1.86 \times 10^{-3} T^{3/2}}{P \sigma_{AB}^2 \Omega_D} \left(\frac{1}{M_A} + \frac{1}{M_B} \right)^{1/2} \tag{7}$$

where M_A and M_B = molecular weights, T = temperature in K, P = gas pressure in atmospheres, σ_{AB} = minimum distance between the centers of two molecules, in angstroms, and Ω_D = dimensionless functions of temperature and intermolecular potential field for one molecule of A and one molecule of B.

A practical method of determining binary gaseous diffusion was developed by Giddings and coworkers,[4] who proposed the following equation by correlating a large number of diffusivity data:

$$D_{AB} = \frac{1.00 \times 10^{-3} T^{1.75}}{P[(\Sigma v_i)A^{1/3} + (\Sigma v_i)B^{1/3}]} \left(\frac{1}{M_A} + \frac{1}{M_B} \right)^{1/2} \tag{8}$$

where v_i is the empirical atomic diffusion volumes.

These atomic diffusion volumes have to be added together for the molecules of both components to give the pertinent values of molecular volumes. These

TABLE I Diffusivities in Gases at 1 atm

System	Temperature (K)	Diff. Coeff. (cm²/s)
Hydrogen–nitrogen	200	0.401
Hydrogen–nitrogen	273	0.708
Hydrogen–nitrogen	400	1.27
Hydrogen–water	307	1.020
Hydrogen–*n*-hexane	289	0.290
Carbon dioxide–water	329	0.211
Air–water	313	0.277

Note: Adapted from Reference 4.

equations are sound at low pressures, where the diffusivities are essentially independent of concentration.

B. Diffusion in Liquids

Solute diffusivities in liquids are ordinarily strongly contingent on the composition, and their magnitude is resolved primarily by solvent properties. The diffusion coefficients in liquids (Table 2) are generally smaller than in gases by at least four orders of magnitude.

The Wilke–Chang equation is a well-known example of the various equations that have been proposed for use in predicting diffusivities in liquids.

$$D_{AB} = 7.4 \times 10^{-8} \frac{(\Psi_B M_B)^{1/2} T}{\eta V_A^{0.6}} \tag{9}$$

where V_A = molar volume of solute A, cubic centimeter per gram mole as a liquid at its normal boiling point, M_B = molecular weight of solvent, T = temperature in K, η = viscosity of solution in centipoises, and Ψ_B = "association factor" for the solvent.

The recommended values for Ψ_B are 1.0 for solvents such as ether, aliphatic hydrocarbons, and benzene; 1.5 for ethanol; and 2.6 for water. Equation (9) provides values within 10% of actual values for dilute solutions of nondissociating small- and medium-size molecules.

The solute diffusivity depends strongly on solvent viscosity; for example, see Eq. (9). While the square roots of the molecular weights of common solvents vary slightly, the reciprocal velocities vary several thousand-fold, thus having a large effect on diffusivity. The dependence of D_{AB} on the solute properties is less obvious for small solutes, as exhibited by the 0.6 power of the molar volume of the solute. Table 3 demonstrates clearly that the diffusivities of large molecules and macromolecules are greatly reduced.

The tendency of cations and anions to migrate with different speeds may also affect the diffusivity of electrolytes. A so-called diffusion potential is

TABLE 2 Diffusivities in Liquids for Dilute Solutions at 20 °C

Solute (A)	Solvent (B)	D_{AB} (cm^2/s × 10^5)
Acetic acid	Water	0.88
Acetic acid	Benzene	1.92
Phenol	Water	0.84
Phenol	Ethanol	0.8
Phenol	Benzene	1.54
Sucrose	Water	0.45
Sodium chloride	Water	1.35

Note: Adapted from Reference 1.

TABLE 3 Molecular Weight and Diffusivity

Molecular weight	Diffusivity (cm²/s × 10⁵)	Molecular diameter (A)
10	2.20	2.9
100	0.70	6.2
1000	0.25	13.2
10,000	0.11	28.5
100,000	0.05	62.0
1,000,000	0.025	132

Note: Adapted from Reference 5.

produced, which decelerates the motion of the fast-moving ion and accelerates the slow-moving ion until both are moving at equal velocities. The diffusion coefficient for a binary electrolyte that ionizes into β_+ cations and β_- anions in dilute solutions is given by Eq. (10).

$$D = \frac{RT}{\mathcal{F}} \frac{\beta_+ + \beta_-}{\beta_+ Z_+} \frac{U_k U_a}{U_k + U_a} \quad (10)$$

where U_k and U_a = the mobilities of cations and anions, respectively, $\mathcal{F}$ = Faraday number, and Z_+ = electrochemical valence of the cation.

C. Diffusion in Solids

The diffusion rates in solids are much smaller than in gases or liquids. As a result, separation processes that require mass transfer in solids are of little interest. However, diffusion in porous solid materials, such as ion-exchange resins and membranes, is of great interest in many separation processes. It is a good idea to remember that the solute diffusion is confined mostly to the pore fluid or to the internal surface in these porous media.

V. MASS TRANSFER

Mass transfer through interfaces is very meaningful, and it is the rate-determining step in many separation processes. As a result, the overall efficiency of the separation process is frequently limited by mass transfer. Therefore, it is not surprising that mass transfer through gas–liquid and liquid–liquid interfaces is of great interest to separation scientists.

Let us think about the transfer of a solute from phase X into phase Y. It is presumed that solute concentrations in the bulk phase are constant, C_X and C_Y, but differ at the interface, C_{X_s}, C_{Y_s}. The difference in concentration of $C_X - C_{X_s}$ and $C_Y - C_{Y_s}$ provides the essential driving forces for the solute to diffuse from the bulk phase X to the interface, and from the interface to the bulk phase Y, respectively.

The flux of the solute through the interface, J_i, may be described as follows:

$$J_i = k_X(C_X - C_{X_s}) = k_Y(C_{Y_s} - C_Y) \qquad (11)$$

where k_X and k_Y are the mass transfer coefficients for phases X and Y, respectively.

The equilibrium concentrations may be used for the interface concentrations if it can be assumed that complete equilibrium has been attained at the interface.

A. Flow Through Packed Columns

Packed columns are commonly used in separation processes such as distillation and chromatography. The packed columns can be looked upon as packed beds and are sometimes called by that name. Chromatography in packed columns is the most important example of packed-bed flow. The fluids flowing through the column can be gas, liquid, or a supercritical fluid (see Section III for the differences in diffusion coefficients of these fluids). A similar flow can be found in porous membranes used for membrane separations.

The flow in a column can be compared with the flow of a stream full of boulders. The current changes directions to find open channels between the rocks. The tortuous flow through a column has been depicted in Figure 2.

The structure of packing in these columns is usually quite complicated, and it is not easy to set up a rigorous theoretical treatment. However, the discussion that follows attempts to improve our understanding of packing structure.

In a well-packed column, the diversity of channel diameter and velocity in individual channels is small, so that, as a whole, the column may be treated as a bundle of tortuous capillary tubes. Darcy's law governs the flow of fluids through packed beds (or columns) that are frequently employed in separation processes such as chromatography.

1. Darcy's Law

This law states that the flow velocity is proportional to the pressure gradient:

$$v_0 = B^0 \frac{P_o - P_i}{\eta L} \qquad (12)$$

where B^0 = specific permeability coefficient, v_0 = superficial velocity, P_o and P_i = the outlet and inlet pressures, respectively, L = length of the tube, and η = viscosity.

Superficial velocity (the average linear velocity of the fluid in the column if no packing were present) may be calculated by dividing the flow rate by the cross-sectional area of the empty tube.

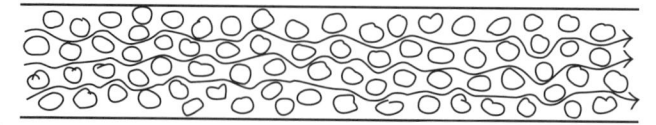

FIGURE 2 Flow through a packed column.

A number of equations have been derived to relate the specific permeability to the particle diameter and bed porosity. The *Kozeny–Carman equation* is the best known of these expressions, and it gives specific permeability as

$$B^0 = \frac{d_p^2 \varepsilon^2}{180(1-\varepsilon)^2} \tag{13}$$

where B^0 = specific permeability coefficient, ε = interparticle porosity, and d_p = particle diameter.

In packed columns, the transition from laminary to turbulent flow is not well defined. However, it is generally agreed that uniform profiles may be obtained when the beds are packed carefully with spherical particles of equal size.

REFERENCES

1. Karger, B. L., Snyder, L. R., and Horvath, C. *An Introduction to Separation Science*, Wiley, New York, NY, 1973.
2. Crank, J. *The Mathematics of Diffusion*, Oxford University Press, London, 1956.
3. Giddings, J. C. *Unified Separation Science*, Wiley, New York, NY, 1991.
4. Fuller, E. N., Schettler, P. D., and Giddings, J. C. *Ind. Eng. Chem.*, 58:19, 1966.
5. *Chemical Engineering Handbook* (J. H. Perry, Ed.), McGraw-Hill, New York, NY, 1963.

QUESTIONS FOR REVIEW

1. Define diffusion and bulk motion. Discuss similarities and differences.
2. Provide equations for Fick's first and second laws.
3. Discuss the usefulness of the Wilke–Chang equation for predicting diffusivity in liquids.

6
CHROMATOGRAPHIC METHODS

I. GENERAL CLASSIFICATION OF CHROMATOGRAPHIC METHODS
   A. Column Chromatography
   B. Paper Chromatography
   C. Thin-Layer Chromatography
   D. Gas Chromatography
   E. High-Pressure Liquid Chromatography
   F. Ion-Exchange Chromatography
   G. Gel Filtration Chromatography
   H. Supercritical Fluid Chromatography
II. CLASSIFICATION BASED ON RETENTION MODE
   A. Enclosed Versus Open-Bed Chromatography
III. CLASSIFICATION BASED ON SAMPLE INTRODUCTION
   A. Frontal Chromatography
   B. Displacement Chromatography
   C. Elution Chromatography
IV. SEPARATION CHARACTERISTICS OF CHROMATOGRAPHIC METHODS
   A. Sample Type
   B. Load Capacity
   C. Fraction Capacity
   D. Selectivity
   E. Speed of Analysis
V. BASIC CHROMATOGRAPHIC THEORY
REFERENCES
QUESTIONS FOR REVIEW

It may be recalled from the discussion in Chapter 1 that column chromatography, where gravity is the main driving force for moving the solvent down the column, is the oldest form of the various chromatographic techniques developed to achieve a desired separation. Its origin can be traced back to the discovery by Tswett early in the twentieth century. He published a description of the separation of chlorophyll and other pigments in a plant extract in 1906, using an apparatus similar to the one shown in Figure 1.

Tswett placed the sample solution in petroleum ether at the top of a column. The column was simply a narrow glass tube packed with powdered calcium carbonate. When the sample was "washed" through the column (the more appropriate term is eluted) with additional petroleum ether, he observed that pigments were separated (resolved) into various zones. After carefully removing the packing, he was able to extract the individual constituent from the zones

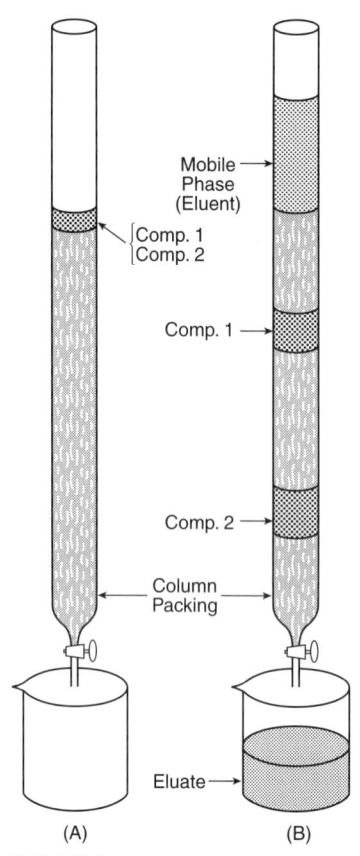

FIGURE 1 Apparatus for column chromatography.

and identify each of them. He explained the observed separation as follows: More strongly retained components "displaced" the more weakly adsorbed ones.

The appearance of the colored bands (zones) led to the name "chromatography" from the Greek words for *color* and *to write* for the above-mentioned separation. It is quite clear that this brilliant discovery in separation science was made because the sample contained colored pigments. However, it should be recognized that the coincidental use of colored compounds was serendipitous and that the color itself had very little effect, if any, on this separation. Unquestionably, the chemical structures and the resulting physicochemical properties of the colored compounds played a significant role in this chromatographic separation.

Let us now recall from the discussion in earlier chapters of this book that the separations achieved by chromatography require two phases, where one phase is held immobile and is called stationary phase, and the other phase, the mobile phase, is allowed to pass over it. In principle, chromatography is similar to the distribution processes discussed in Chapter 2; more specifically, the Craig multiple extraction process. Both utilize two phases, one of which moves with respect to the other. The primary point of interest is that in chromatography the mobile phase moves continuously past the stationary phase. In the above-mentioned example of column chromatography, the equilibration between the phases is very fast; however, it should be recognized that the equilibrium is never complete at each point on the column. A significant advantage of chromatography is that a variety of phases can be used, whereas Craig's process is limited to the use of *two* immiscible solvents.

A number of chromatographic methods are available to achieve separation of various samples. The great importance achieved by chromatography is based on its speed, resolving power, and ability to handle a small amount of sample.[1,2] A good workable theory has been developed that provides the basis of separations achieved by chromatography and how these separations can be optimized. In addition, the technique and equipment used are reasonably simple and easy to operate. All these factors have resulted in a numerous applications.

An ultimate in separation simplicity has been carried out with the techniques of paper and thin-layer chromatography, where separations are achieved on a piece of paper or a layer of sorbent on a plate (see Chapters 7 and 8).

Gas chromatography is discussed in Chapter 9. A photograph of a gas chromatograph is shown in Figure 2. The speed of gas chromatographic analysis is shown in Figure 3. The complete natural gas analysis can be performed in 90 s with a 4-channel gas chromatographic system.

A photograph of GC/MS instrument is shown in Figure 4. Figure 5 demonstrates extraordinary selectivity offered by GC, as exemplified by the separation of chlorinated pesticides at the picogram level. A very impressive

FIGURE 2 Gas chromatograph (Courtesy Shimadzu Instruments).

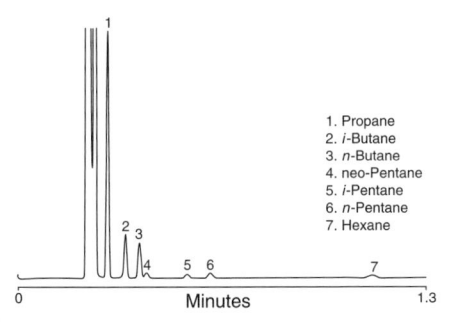

FIGURE 3 Natural gas analysis on Channel A: CP-Sil 5 CB (Courtesy Varian Instruments).

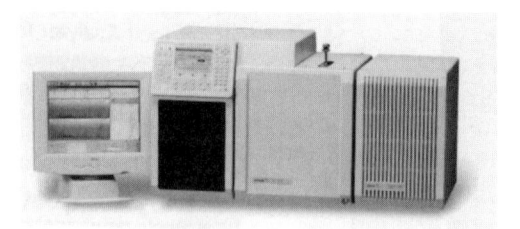

FIGURE 4 GC/MS instrument (Courtesy Varian Instruments).

Rapid-MS. Chromatogram of Chlorinated Pesticides (pg level)

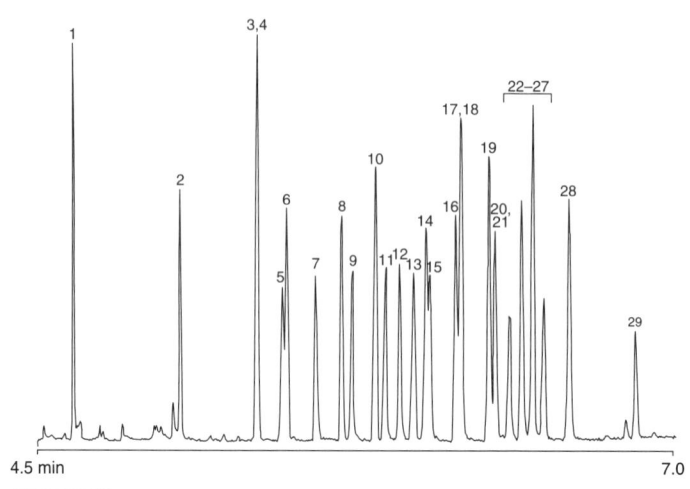

FIGURE 5 GC/MS chromatogram of chlorinated pesticides (Courtesy Varian Instruments). (1) HCB. (2) Benzene, pentachloro-. (3) Benzene, hexachloro-. (4) α-Lindane. (5) β-Lindane. (6) γ-Lindane. (7) δ-Lindane. (8) PCB28. (9) Heptachlor. (10) PCB52. (11) Aldrin-R. (12) Isobenzan. (13) Isodrin. (14) Heptachlor epoxide. (15) Heptachlor epoxide. (16) o,p'-DDE. (17) PCB101. (18) Endosulfan. (19) Dieldrin. (20) p,p'-DDE. (21) o,p'-DDD. (22) Endrin. (23) PCB118. (24) Endosulfan II. (25) p,p'-DDD. (26) p,p'-DDT. (27) PCB153. (28) PCB138. (29) PCB180.

example of selectivity of separation of enantiomeric amino acids can be seen in Figure 13 of Chapter 11.

As mentioned before, in chromatography sample components undergo an equilibrium distribution between two phases. It is this equilibrium that controls the velocity with which each component migrates through the system. Band broadening or dispersion of each component in the direction of migration also takes place. These two phenomena, band broadening and differential migration, determine the extent of separation of the starting sample.

I. GENERAL CLASSIFICATION OF CHROMATOGRAPHIC METHODS

Chromatographic methods may be classified as follows:[3]

- Column chromatography
- Paper chromatography

- Thin-layer chromatography
- Gas chromatography
- High pressure liquid chromatography
- Ion-exchange chromatography
- Gel filtration chromatography
- Supercritical fluid chromatography

A short introduction of the various chromatographic methods follows to familiarize the reader with the diversity of available methods.

A. Column Chromatography

An introduction to this technique was provided above. It may be recalled that Tswett used column chromatography when he made his brilliant discovery of chromatography. This technique is further covered in some detail in Section II. A.

B. Paper Chromatography

Paper chromatography is generally carried out on strips of filter paper, where the sample is applied a few centimeters from one end and then the strip is developed in an ascending or descending mode in a saturated chamber with a solvent or mixture of solvents. The descending mode is generally the preferred technique. Separations are frequently carried out for a long time, such as overnight, to provide higher efficiency (the longer development distance on the paper strip provides a larger number of theoretical plates). The strips are dried and the components are detected visually or by some other suitable technique (see Chapter 7).

C. Thin-Layer Chromatography

Thin-layer chromatography is performed by coating a glass plate with a thin layer of silica gel. The coating is roughly 250 µm thick. Thicker coatings are applied for preparative separations. The plate is developed in a closed chamber that is generally saturated with the developing solvent(s). The plates can be developed horizontally or in an ascending or descending mode. The ascending mode is the most commonly used type. Following development to a given distance, say 10 cm, the plates are dried and the sample is detected visually or by various other techniques (see Chapter 8).

D. Gas Chromatography

The type of mobile phase used, gas or liquid, leads to the general terms gas chromatography (GC) or liquid chromatography (LC). In gas–liquid chromatography (GLC, or more commonly, GC) the stationary phase is a nonvolatile liquid (at the operating temperature) coated onto a porous support or onto the walls of a capillary tube. The mobile phase is an inert carrier gas, for example, helium. Separation occurs as a result of the differences in vapor pressure above the liquid phase of the various sample components–compounds with higher vapor pressures progress through the bed at a more rapid pace.

Gas chromatography entails those separations where one of the phases is a gas and the other (stationary phase) allows adsorption or partition. The sample is detected by nondestructive detectors such as thermal conductivity or destructive detectors, as exemplified by the flame ionization detector (see Chapter 9).

Ion-exchange and gel permeation chromatography have also been carried out with this technique.

E. High-Pressure Liquid Chromatography

High-pressure liquid chromatography (HPLC, or more commonly, LC) involves separation on a column under pressure. The columns commonly used for HPLC may contain an adsorbent or materials that permit partition, ion exchange, or molecular permeation. Pressures up to 6000 psi may be used. The mobile phase may be composed of one solvent or a mixture of solvents, with or without modifiers. Detection is commonly performed by UV; however, a number of other detectors have been used (see Chapter 10 for more details).

F. Ion-Exchange Chromatography

This technique has flourished as a specialized field of liquid–solid chromatography that is suitable for ionic species. The stationary phase is especially designed so that it allows exchange of cations or anions. These separations can be carried out in a gravity-fed column or a TLC plate or at high pressure with an HPLC instrument.

G. Gel Filtration Chromatography

A separation process carried out with a gel consisting of modified dextran (a three-dimensional network of linear polysaccharide molecules that have been cross-linked). The material swells in aqueous solutions and produces a sieve-like structure that can separate molecules by their size. As a result, this type of separation is often called size-exclusion chromatography (see Section II). These separations may also be carried out at high pressure with an HPLC instrument.

H. Supercritical Fluid Chromatography

Supercritical fluid chromatography (SFC) involves utilizing a gas such as carbon dioxide at a certain pressure and temperature so as to yield a supercritical fluid. Certain modifiers may be added to improve separations. This technique provides the advantages of both gas chromatography, in terms of detectors, and HPLC, in terms of selectivity improvement via the mobile phase.

Detailed discussions on paper chromatography, thin-layer chromatography, gas chromatography, and high-pressure liquid chromatography may be found in Chapters 7–10. Chapter 11 covers supercritical fluid chromatography (SFC) as well as a number of other separation techniques including capillary electrophoresis (CE), and it provides a strategy for selecting a separation method.

II. CLASSIFICATION BASED ON RETENTION MODE

Chromatographic methods classification may be done on the basis of the mode of retention in the stationary phase:

- Sorption
- Exclusion
- Ion exchange

1. Sorption

Sorption is a general term used to denote both adsorption and partition; that is, when it is not clear that only one of these mechanism is involved and indications are that both mechanism play a role. Chromatographic separation based on only adsorption or partition is possible, and the method is called adsorption or partition chromatography accordingly; however, mixed-mode separations are often encountered in chromatography.

2. Exclusion

Exclusion relates to separation based on the size of the sample molecule. An exclusion mode of chromatographic separation does not allow the sample to enter the stationary phase based on size or shape; it forms the basis of gel permeation chromatography or size-exclusion chromatography.

3. Ion Exchange

Ion exchange, as explained previously, relates to the exchange of cations or anions. In some cases the molecules may be made ionic by the appropriate use of pH adjustment (see discussion in Chapter 2). For example, amines can be protonated based on their pK_a to yield cations in the acidic pH range.

A. Enclosed Versus Open-Bed Chromatography

Sorption, exclusion, and ion-exchange chromatography may be carried out in an enclosed bed, for example a column, or in an open bed such as a thin-layer plate coated with an appropriate stationary phase.

1. Column Chromatography

Liquid–solid adsorption chromatography entails packing a column with some adsorbent such as alumina or silica gel, and the applied sample is eluted with a nonpolar solvent by gravity flow through the column. The technique can be best described as gravity-fed adsorption column chromatography, where gravitational force is driving the solvent down and through the column.

The liquid–solid adsorption process may be described as follows.[3] The properties of molecules or ions located at the surface of a solid particle are somewhat different from the properties of the same species in the interior. The bonds within the surface layer are disturbed by the lack of overlaying structure. Therefore, the surface layer is at a high energy level and is characterized by its surface activity. The active surface attracts and tends to adsorb species from the fluid (gas or liquid) if a solid particle is immersed in a fluid. The attractive forces discussed earlier in Chapter 4 can be electrostatic, dipole–dipole, dipole-induced dipole, London forces, or a combination of several of these forces. When a solution is flowing through the column, any or all of the solutes as well as the solvent may be adsorbed on the column.

A good adsorbent should have a large surface area with a large number of active sites on the surface. The surface activity is largely lost when it is covered by a monolayer of the adsorbed species. When a solution flows over the surface of an adsorbent, an equilibrium is achieved for the adsorption or desorption of the species present. The affinity between the concentration of a given species in a solution and the quantity that is adsorbed may be plotted, and the line thus obtained is specified as the adsorption isotherm. Such an isotherm may be linear, convex, or concave (see Chapter 3). The effect of these isotherms on peak shapes observed in chromatography is discussed next.

2. Peak Tailing/Leading

The rate of travel of a component through a chromatographic system is a function of the fraction of the component found in the mobile phase. Each component leaves the column in the form of a band or peak, which is generally symmetrical and bell-shaped (a Gaussian, or standard-error, curve). The band spread leads to a region of high concentration in the center of the band relative to the edges. If the isotherm is convex, the center of the band travels at a faster rate than the leading and trailing edges. In the extreme case, the center will nearly overtake the leading edge, giving a very sharp front and a long trailing edge. This is called peak tailing and is not desirable in chromatography. The opposite case gives a long leading edge and a sharp back to the peak; this is called peak leading.

Peak tailing is more pronounced when active adsorbents are involved. One way to reduce this effect is to partially deactivate the solid by covering the most active sites with another substance or by raising the temperature. The reduction of sample size may also help minimize peak tailing in that the system is allowed to remain within the linear portion of the isotherm at very low concentrations.

3. Adsorbents

Silica gel and alumina are commonly used adsorbents. The adsorbing power of the material depends on the chemical nature of the surface, the area available, and its pretreatment. Some common adsorbents are listed here in the order of decreasing adsorbing power.

- Alumina
- Charcoal

- Silica gel
- Calcium carbonate
- Starch
- Cellulose

The adsorptive properties can be easily modified; however, they are difficult to control, so the reproducibility of surfaces is a common problem in adsorption chromatography. The surface activity of the common adsorbents varies from site to site on the same particle, and from batch to batch. Pretreatment, including exactly prescribed washing and drying conditions, is very desirable.

4. Eluting Solvents

The choice of the eluting solvent is almost as important as the choice of adsorbent. The liquid mobile phase composed of one or more solvents provides transportation to the sample as a carrier, but more importantly, it influences the distribution coefficient through its solvent power. In addition to the relative solubilities of the solutes in the eluting solvent, it is also necessary to consider the competition between the solutes and solvent for the adsorption sites on the surface of the stationary phase. A solvent that elutes the solutes very rapidly is not likely to separate them very well. On the other hand, a solvent that elutes components too slowly is likely to give long retention times. Long retention times frequently lead to excessive band broadening and unnecessary dilution of the sample.

Some commonly used solvents in the order of increasing eluting power on alumina are shown here.

- Petroleum ether
- Carbon tetrachloride
- Isopropyl ether
- Toluene
- Chloroform
- Diethyl ether
- Methyl ethyl ketone
- Acetone
- Ethyl acetate
- Dioxane
- Acetonitrile
- Ethanol
- Methanol
- Water

As discussed earlier, adsorption involves interaction or fixation of sample components at a surface or with fixed sites. More than one retention mode can also operate in a given system. It may be useful to deliberately create a dual mode of retention to enhance differences in migration of the sample components.

Adsorption chromatography is called liquid–solid chromatography (LSC) when one of the phases is liquid. It is based on a solid (usually porous) stationary phase and a liquid mobile phase. This technique is currently restricted to the separation of quite volatile samples and the characterization of solid surfaces.

Column chromatography offers greater sample capacity; that is, a larger amount of sample can be applied to the column. The technique is preferred for preparative separations, repetitive quantitative analyses, and difficult separations. Column chromatography may be additionally subdivided based on whether the packed column contains adsorbent or other materials that lead to partition, ion exchange, and so forth. Partition, ion exchange, or separations based on molecular weight can thus be carried out, depending on the packing in the column. Coated capillary columns may also be used. In column separations, mobile phase flow is achieved by a pressure drop along the column; however, in open-bed chromatography this flow is the result of capillary wetting.

5. Open-Bed Chromatography

When no column is used and the stationary phase is open, the technique is called open-bed chromatography or planar chromatography. Examples include paper chromatography and thin-layer chromatography (see the earlier description and also refer to Chapters 7 and 8). Open-bed chromatography has the advantages of simplicity and flexibility and is most useful for qualitative analysis and the empirical selection of mobile and stationary phases for a given separation. Various separation mechanisms can be employed, such as adsorption, partition, or ion exchange.

III. CLASSIFICATION BASED ON SAMPLE INTRODUCTION

Another method of classification is based on the manner in which the sample is introduced into the bed and migrates through it. This includes the following chromatographic techniques:

- Frontal chromatography
- Displacement chromatography
- Elution chromatography

A. Frontal Chromatography

In frontal chromatography, the sample is introduced into the bed continuously, rather than in small portions. Here the sample itself constitutes the mobile phase. The sample components are selectively retarded; and fronts, rather than bands, are formed as a result of the separation. The less retained component emerges from the column first, followed by the mixture of the first component plus the next most strongly retained component. Frontal displacement cannot accomplish complete recovery of the pure sample; however, it is useful for concentrating trace impurities and for purifying large volumes of liquids (e.g., in the deionization of water).

B. Displacement Chromatography

In displacement chromatography, the mobile phase is much more strongly retained by the stationary phase as compared with the sample. In this situation, one can consider that the sample on the stationary phase has been replaced by the mobile phase. Other displacers, that is, compounds that are retained more strongly than the sample, can also be used. The main advantage of displacement chromatography is that greater loads of sample may be applied to the bed. Therefore, it is useful for preparative- or production-scale samples. The main difficulty of utilizing this technique is to find the right displacer. Displacement chromatography has been found to be very useful in bioseparations of proteins.[4]

C. Elution Chromatography

In elution chromatography, sample components migrate much more slowly than the mobile phase, so that adjacent bands are usually separated by the mobile phase. Elution chromatography is used almost exclusively for analytical separations and is carried out under fixed or constant separation conditions (constant temperature and mobile phase flow rate with the same mobile and stationary phases throughout the separation). However, at times it is advantageous to change experimental conditions systematically; for example, temperature programming, flow programming, or gradient elution may be changed at a predetermined rate.

Elution separations may be further subdivided into nonlinear and linear elution based on the shape of the isotherm under the conditions of separations.

1. Nonlinear Elution Chromatography

Nonlinear elution is usually observed when high sample concentrations are used, for example, in preparative separations. Gaussian bands are generally not obtained in nonlinear elution chromatography because high sample concentrations are frequently used. Depending upon the nature of isotherm, the peaks either tail as is the case with convex isotherm or lead in the case of concave isotherm. The point to remember here is that when K varies with sample size, elution bands become asymmetrical, nonlinear elution occurs, a poor separation is obtained, and it is difficult to identify compounds based on their migration rate. As a result, most analytical and preparative separations are carried out under linear isotherm conditions.

2. Linear Elution Chromatography

In linear elution chromatography, the distribution constant, K, is independent of sample concentration, and the band is symmetrical. It should be clear from earlier discussion that differential migration (or retardation) rate of the various sample components (solutes) has an important effect on their chromatographic separation. When two compounds migrate at the same speed, no separation is observed. A typical chromatogram showing the separation of two compounds is shown in Figure 6.

The fundamental retention parameter in column chromatography is the retention volume, V_R, defined as the volume of the mobile phase that must flow

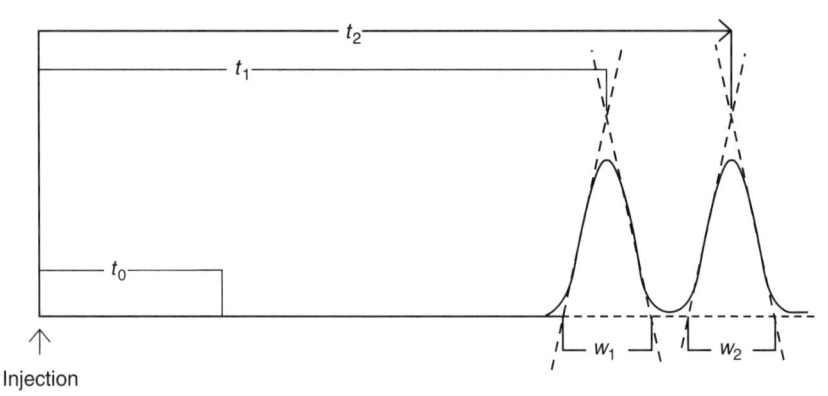

FIGURE 6 A typical chromatogram.

through the column for elution of a particular component. The retention time, t_R, required for elution of a band (or solute) may also be obtained from V_R and the volumetric flow rate F (mL/min) of the mobile phase:

$$t_R = V_R/F \tag{1}$$

The time for elution of an unretained component is t_0.

At the band center, the distribution of solute molecules between stationary and mobile phases is approximately at equilibrium. We can relate t_R to the equilibrium properties of the system. A useful measure of retention (R) is the fraction of solute in the mobile phase or the probability that a solute molecule will be detected in the mobile phase at any given instant. We may signify R as a function of the total moles of solute in the stationary and mobile phases, n_S and n_M, respectively.

$$R = \frac{n_M}{n_M + n_S} \tag{2}$$

Because the capacity factor, k', has been characterized as the ratio of total solute distributed between two phases (n_S/n_M), we find that

$$R = \frac{1}{1 + k'} \tag{3}$$

The term k' can be related to the distribution constant, K,

$$k' = KV_S/V_M \tag{4}$$

where V_S is the volume of the stationary phase, V_M is the volume of the mobile phase.

Since K is constant for linear elution, R is also constant. The larger the value of R, the faster a solute moves through the column. The average migration velocity, v_S, of the solute is equal to mobile phase velocity, v, times the fraction of solute R in the mobile phase at a given time.

$$v_S = vR \tag{5}$$

When R is zero, the solute does not move at all and thus $v_S = 0$, as determined by the above equation. And when $R = 1$, the solute moves with the same velocity as the mobile phase and $v_S = v$.

The migration rate of sample or solute retention in open-bed chromatography is measured in terms of the R_f value: the distance migrated by the solute, divided by the distance migrated by the mobile phase (solvent front). The R_f value is proportional to the velocity ratio v_S/v, from which we obtain $R_f = R$, or it can be related to the capacity factor (k') as follows:

$$R_f = \frac{1}{(1 + k')} \tag{6}$$

We can define another useful parameter, R_M, that can be related to the logarithm of the capacity factor as follows:

$$R_M = \log\left(\frac{1 - R_f}{R_f}\right) \tag{7}$$

$$= \log k' \tag{8}$$

IV. SEPARATION CHARACTERISTICS OF CHROMATOGRAPHIC METHODS

There are a number of significant characteristics of various chromatographic methods that are relevant to the separation goal and can serve as the basis for selection of a given method.[1] The important characteristics are listed here.

- Sample type
- Load capacity
- Fraction capacity
- Selectivity
- Speed of analysis

In Table 1, a number of chromatographic methods for separating chemical compounds are compared with respect to the features listed.

TABLE I Characteristics of Various Chromatographic Methods

Method	Sample type	Selectivity	Fraction capacity	Load capacity	Speed
Column	Nonvolatile	Phys./isomer	10–100	mg–g	Slow
Paper	Nonvolatile	Phys./chem.	10	µg–mg	Slow
TLC	Nonvolatile	Phys./isomer	10–50	mg–g	Fair
GC	Volatile	Mol. wt.	100+	mg–g	Fast
HPLC	Nonvolatile	Phys./chem.	10–50	mg–g	Fast
SFC	Nonvolatile	Phys./chem.	10–50	mg–g	Fast

Note: Adapted from Reference 1.
Phys. = physical interactions; chem. = functional groups interactions; mol. wt. = molecular weight.

A. Sample Type

The adaptability of a separation method relates to its ability to be used with certain types of samples with special consideration regarding their physicochemical properties. Materials within the range of conventional separation methods may be classified as atomic, small molecules, macromolecules, or particulates. Small molecules (with molecular weight less than 2000) and macromolecules (with molecular weight greater than 2000) are of major interest to us. Macromolecules not only are nonvolatile but also possess some properties that necessitate the use of separation processes having appropriate adaptability. Examples of such chromatographic and nonchromatographic separation methods are gel chromatography and ultracentrifugation, respectively. In some instances, particularly in analytical work by GC, it is advisable to make nonvolatile molecules volatile by derivatization.

B. Load Capacity

The term load capacity deals with the maximum quantity of a mixture that can be separated with a given efficiency by a particular process. For example, distillation and crystallization may be carried out easily on a production scale, but they are less suitable for the separation of a few micrograms in analytical work because load capacity of these methods is limited to a high amount of sample. Chromatography excels in this area because it is useful for separating μg to kg quantities of materials.

C. Fraction Capacity

The fraction capacity of a separation process may be defined as the maximum number of components that can be separated in a single operation. In chromatography, fraction capacity is equivalent to peak capacity, which is the maximum number of separable bands or compounds yielded by a chromatographic system for a given set of experimental conditions. In electrophoresis, chromatography, and mass spectrometry, the fraction capacity may be relatively high, normally between 10 and 100. Fraction capacity is related to the efficiency of a technique for separating sample mixtures in a single step.

D. Selectivity

As mentioned in Chapter 1, the selectivity for a given separation method refers to the intrinsic capability to distinguish between two components. Selectivity is measured by the separation factor, α. Thus the selectivity of a given separation method relates to the ability of that method to discriminate on the basis of one or more properties of the components at the molecular level. Ion-exchange processes have distinct acid or base selectivity. The molecular properties of sample components may cover various selectivity categories:

- Molecular weight or size: homologs within a series, polymers including biopolymers

- Functional groups: carbonyl, hydroxyl, esters, etc.
- pK_a: acidic or basic properties
- Complex formation: metal complexes
- Shape: *n*-hexane versus cyclohexane
- Isomers: geometric, optical, etc.

E. Speed of Analysis

Both speed of analysis and convenience are especially important considerations in those methods that are performed routinely. These factors are interrelated with fraction capacity, load capacity, and efficiency. The first two are generally accomplished by sacrificing efficiency and time.

V. BASIC CHROMATOGRAPHIC THEORY

The basic chromatographic theory can be credited to the work of Martin and Synge on liquid partition chromatography. Their work provided the basis for a more refined and detailed treatment of GLC. Discussed here are theoretical principles that apply to all types of chromatography.

In a liquid–liquid partition process, a sample molecule may be present in the mobile phase or stationary phase. When it is in the mobile phase, it is moving at the same rate as the mobile phase; however, when it is in the stationary phase, it makes no forward progress.

The distribution of sample between the two phases is governed by an equilibrium constant, known as distribution coefficient or partition coefficient, K. Equilibration is a dynamic process, and a given molecule passes back and forth between the two phases so that, on average, the concentrations obey the distribution law:

$$K = C_S/C_M \tag{9}$$

where C is concentration, C_S and C_M are concentrations in the stationary and mobile phases, respectively.

1. Capacity Factor

Let us elaborate further on our discussion of capacity factor as it relates to chromatography (you may want to review the information provided earlier in this chapter as well as that given in Chapter 3). The distribution coefficient K, as defined previously, applies to various mechanisms encountered in chromatography, depending on the nature of the phases and types of interaction of the sample components with each phase. Regardless of whether the process is adsorption, partition, or ion exchange, the value of K determines the relative proportion in the two phases. For a truly dynamic equilibrium, the fraction of total time that an average molecule spends in the mobile phase is directly related to the fraction of total population that is found in the mobile phase.

Fraction of time spent by sample molecules in mobile phase

$$= \frac{\text{Number of molecules in mobile phase}}{\text{Total number of molecules}}$$

$$= \frac{C_M V_M}{C_M V_M + C_S V_S} \tag{10}$$

$$= \frac{1}{1 + K V_S / V_M} \tag{11}$$

$$= \frac{1}{1 + k'} \tag{12}$$

where k' is capacity factor.

Capacity factor (k') is an important parameter in chromatography and is equal to $C_S V_S / C_M V_M$, or the ratio of moles of a sample component in the stationary phase divided by the moles in the mobile phase.

2. Travel Rate of Sample

The fraction of time spent by a sample in the mobile phase is given by Eq. (12) above. If we multiply this equation by the linear velocity, u, of the mobile phase, we get the travel rate of the average sample molecule:

$$\text{Travel rate} = u \left(\frac{1}{1 + k'} \right) \tag{13}$$

The travel rate of an average molecule depends on the following factors:

- Flow rate of the mobile phase
- Ratio of volume of stationary phase to volume of mobile phase
- Distribution coefficient for each component

In chromatographic analysis of a sample containing various components, the sample is injected at the inlet of the column. All sample molecules start from the same place at the same time, but each component moves at a different rate. We can look at the separation in terms of time (e.g., in paper chromatography) or in distance (the outlet of the column). Each component appears after a different interval and is detected by the appropriate detector. A plot of detector response versus time is called a chromatogram. Such a plot resembles a distribution diagram within the column (concentration versus distance at a fixed time), but the two types of plots should not be confused.

3. Retention Time

The time a component takes to travel the length of the column, L, is known as the retention time, t_R. It can be represented in an equation form as follows, based on the rate equation shown above:

$$t_R = \frac{\text{Column length}}{\text{flow rate}}$$

$$= \frac{L}{\mu}(1 + k')$$

$$= t_M (1 + k') \tag{14}$$

where t_M is the time required for the mobile phase to travel the length of the column.

We can calculate the volume required to elute a component by multiplying its retention time by the flow rate of the mobile phase. The separation factor, α, is given by the following equation:

$$\alpha = \frac{t_R - t_M}{t_R^* - t_M} \tag{15}$$

where t_R^* refers to retention time of a reference standard.

4. Theoretical Plates

The similarity between the processes occurring in a chromatographic column and a fractional distillation column most likely led Martin and Synge to adapt the theoretical plate concept that had been successfully used in treating separation by distillation. The similarity of these processes to those occurring in the Craig apparatus discussed in Chapter 2 is noteworthy. A chromatographic column can be assumed to be made up of a large number of identical segments or theoretical plates, where equilibration is achieved in each segment. This model is not exact; however, deviations are not important when a large number of plates are present. In chromatography, we generally use the concentration as a function of time as it reaches the detector at the column outlet. The mathematical relationship between the distribution along the column and the number of transfers or theoretical plates is similar for the two processes.

In chromatography, the total number of theoretical plates, N, in a given column may be determined from the observed chromatogram by measuring the retention time and peak width as shown in Figure 6.

$$N = \frac{(4t_R)^2}{W} \tag{16}$$

Both retention time and peak width must be measured using the same units.

The number of theoretical plates in a column is a function of column preparation, the characteristics of the solute, temperature, flow rate, method of sample introduction, and so forth. As a result, N is only an approximate number that is useful for comparative purposes, and it is frequently described as column efficiency.

A useful measure of column performance is the height equivalent of each theoretical plate, or HETP, or more simply, H. Small values of H are very desirable.

$$H = \frac{L}{N} = \frac{L}{16}(W/t_R)^2 \tag{17}$$

5. Rate Theory

The sample molecules, when injected onto a chromatographic column, travel in a random fashion that is influenced by a number of factors affecting the value of H.

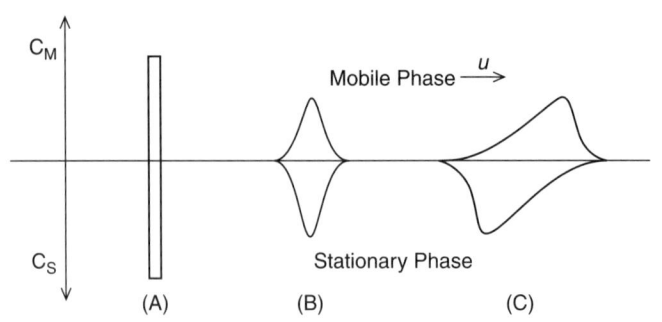

FIGURE 7 Concentration profiles in a chromatographic column.

1. The molecules take random paths; this would be true even if we were using an unpacked column. There may be gross variations in the stream lines if the column is packed with irregularly shaped particles. This means that molecules that started together at the inlet will arrive at the outlet over a period of time.
2. The molecules are inclined to diffuse from the region of high concentration to the region of low concentration. As the zone passes through the column, it is certain that the diffusion will cause both forward and backward spreading as shown in Figure 7.
3. A dynamic equilibrium may be established quickly, but it is generally not instantaneous. The continuous flow of the mobile phase causes a deviation from the equilibrium: The ratio C_S/C_M is smaller than K at the leading edge of a zone and larger than K at the trailing edge. This is illustrated in Figure 7 (C). The zone spreads in both directions as a result of this time lag.

The total effect of these three factors operating independently is given by the following equation:

H = Contributions from unequal paths (eddy diffusion)
  + Contributions from diffusion along column (longitudinal diffusion)
  + Contributions from nonequilibrium (mass transfer)

A mathematical expression for the above statement, known as the van Deemter equation, was first derived by a group of Dutch petroleum scientists:

$$H = A + B/u + Cu \qquad (18)$$

The terms A, B, and C are quite complex and are dealt with later in this book (see Chapter 9). The point to remember is that they are constant for a given column, but each constant includes several experimental parameters specific to each type of chromatography. A few general observations are given here.

1. Small, tightly packed columns give a small A term. In most well-packed columns (Fig. 8), the value of the A term is close to zero and can be neglected.

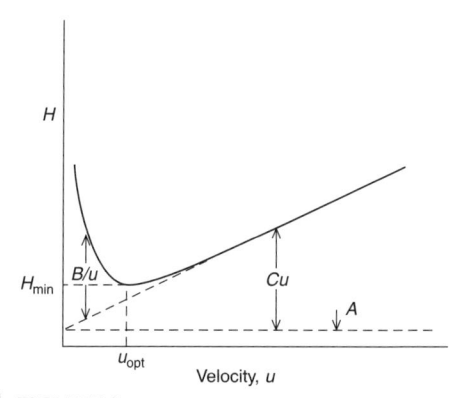

FIGURE 8 Van Deemter plot.

2. The B term is related to diffusion along the length of the column. The diffusion in liquids is approximately 10^5 times slower than that in gases. As a result, B term is less important in a liquid mobile phase. The longitudinal diffusion is influenced inversely by the velocity of mobile phase. At low velocity, the elution time is longer; as a result, there is more time for diffusion to spread the band.
3. The C term is a complex function of the geometry of the phase, distribution coefficient, and the rate of diffusion in both phases. Mass transfer is directly related to the velocity of the mobile phase. At high velocity, there is less time available for equilibration.

For best separations, H should be minimized, which in turn maximizes n and yields narrow peaks. The van Deemter equation given above is based on many approximations, and many workers have modified and expanded it to arrive at a closer fit to experimental data.

6. Resolution

The resolution of two peaks can be determined practically by the following equation:

$$R = \frac{2(t_2 - t_1)}{W_1 + W_2} \tag{19}$$

where R is resolution, t_2 and t_1 are retention time of peaks 2 and 1, respectively, $W_1 + W_2$ is peak width of peak 1 and 2, respectively.

A resolution of 2 gives baseline separations; however, lower resolution values based on practical observations are frequently acceptable for obtaining separations.

REFERENCES

1. Karger, B. L., Snyder, L. R., and Horvath, C. *An Introduction to Separation Science.* Wiley, New York, NY, 1973.
2. Ahuja, S. *Selectivity and Detectability Optimization in HPLC.* Wiley, New York, NY, 1989.

3. Pecosk, R. S., Shields, L. D., Cairns, T., and McWilliam, I. G. *Modern Methods of Chemical Analysis*. Wiley, New York, NY, 1976.
4. Shukla, A. A. and Cramer, S. M. In *Handbook of Bioseparations* (S. Ahuja, Ed.). Academic Press, New York, NY, 2000.

QUESTIONS FOR REVIEW

1. Define chromatography.
2. List various forms of chromatography, and differentiate them from each other by short descriptions.
3. Provide a simple form of the van Deemter equation, and describe its usefulness in chromatographic separations.

7
PAPER CHROMATOGRAPHY

    I. CHROMATOGRAPHY PAPER
       A. Unmodified Cellulose
       B. Modified Cellulose Papers
       C. Glass-Fiber Papers
       D. Membrane Filters
   II. SAMPLE PREPARATION
  III. SAMPLE CLEANUP
       A. Sample Solubility
       B. Removal of Inorganic Materials
       C. Removal of Proteins
       D. Removal of Lipids
  IV. DERIVATIZATION
   V. MOBILE PHASES AND STATIONARY PHASES
       A. Solvents
       B. Stationary Phases
       C. Mobile Phases
  VI. DEVELOPMENT OF CHROMATOGRAMS
 VII. DETECTION
VIII. QUANTITATION
     REFERENCES
     QUESTIONS FOR REVIEW

Paper chromatography, as the term implies, is carried out on paper; for example, an ordinary filter paper can be used for this purpose. Various types of papers that have been used for this technique are given in Section I. The main characteristics of paper chromatography are the flat-bed arrangement and the choice of heterogeneous phases. Both phases are liquid; that is, the technique is based on separation by liquid–liquid chromatography. This is achieved in paper chromatography by immobilizing a liquid (referred to as the stationary phase) on paper. It is important to recognize at this point that paper contains a significant amount of water.

    As mentioned above, paper chromatography is a flat-bed or an open-bed technique that is similar to thin-layer chromatography (discussed in Chapter 8), where a thin layer of sorbent is coated onto a glass plate or foil. Both of these techniques are related in terms of sample application, development, detection, and quantitation.

    The main advantages that paper chromatography offers are simplicity, low cost, and unattended, hassle-free operation. It can be run in various modes, and quantitation may be achieved without the use of expensive instrumentation.

I. CHROMATOGRAPHY PAPER

The fibrous structure rather than the chemical composition characterizes the paper used in paper chromatography. Investigations have revealed that improvements in the fibrous structure can help improve the chromatographic performance. Therefore, attempts have been made to prepare paper that is utilized for chromatography by using a cellulose slurry with very short fibers or to prepare sheets from a slurry of microfilaments of glass fiber impregnated with sorbents.[1]

As mentioned before, regular filter papers are suitable for paper chromatography. The well-known examples are Whatman 1 and Schleicher and Schuell 2043b; they have been used for a variety of separations. A number of modifications have been introduced over the years to meet various needs. Discussed below are the various types of papers that have been used for paper chromatography.

A. Unmodified Cellulose

Most of the paper used for paper chromatography is based on unmodified cellulose. Other examples besides the two mentioned above include Machery-Nagel 260, Munktell Chr 100, Ederdol 202, and Whatman 2. These papers are sold as rectangular or circular sheets. Easy commercial availability of these papers helps obtain reproducible results and save time. Slight changes in the width of paper at the point of entrance of solvent may affect separation. Hence, availability of standardized papers is important. Specially shaped papers are also available (see Fig. 1). These papers have approximately the same weight for a given surface area and offer comparable flow rates.

There is a large diversity of papers made especially for paper chromatography, which offer a variety of options:

- Less dense papers for more rapid flow rates (e.g., Whatman 54 and 4)
- Smooth papers (e.g., Schleicher and Schuell 2040b, Whatman 3 MM)

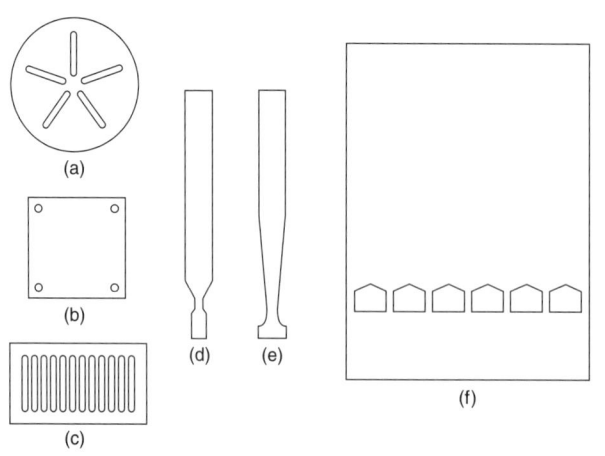

FIGURE 1 Specially shaped papers (1).

7 PAPER CHROMATOGRAPHY

- Acid-washed papers (e.g., Whatman 540-42, Schleicher and Schuell 2040a)
- Papers free of lipid-soluble substances (e.g., Ederol papers labeled with the letter F)
- Preparative papers (e.g., Schleicher and Schuell 2071 and Whatman 31 ET)

In addition, modified cellulose papers that are suitable for chromatographing polar, nonpolar, or ionic substances are available (see the next section).

B. Modified Cellulose Papers

For separations of polar substances, papers with increased capacity have been introduced. For example, carboxyl papers, papers with high carboxyl content (Schleicher and Schuell I or II) are suitable for the separation of polar compounds such as amines, amino acids, and cations.

Papers containing ion-exchange resins offer the possibilities for ion exchange. Examples include cellulose cation- and anion-exchange papers from various manufacturers mentioned above, such as Schleicher and Schuell and Ederol.

For separation of some hydrophobic substances, kieselguhr filter paper (e.g., Schleicher and Schuell 287) has been used. For these substances, papers prepared from cellulose esters, for example, acetylated papers (Schleicher and Schuell 2043b/6 or Machery-Nagel 214 AC) or silicone-treated papers (e.g., silicone oil-impregnated Whatman 1, Machery–Nagel 212, or Schleicher and Schuell 2043b) are used. Other papers containing adsorbents like alumina or silica gel have also been used.

C. Glass-Fiber Papers

These papers have been used, in appropriate applications, because they offer the following advantages:

- Reagents that are too corrosive for cellulose papers can be used.
- Adsorption effects are minimized in some cases.
- Analysis time is reduced significantly.

D. Membrane Filters

Nitrocellulose membranes have been used as carrier materials in the separation of macromolecules. This technique utilizes the fact that high molecular weight materials form a homogeneous immobilized film on the surface of the microporous carrier that is capable of specific interactions with various substances. Sorption can be minimized by utilizing neutral detergents.

II. SAMPLE PREPARATION

To apply a sample on the chromatography paper for development, it is necessary to solubilize solid samples in a solvent that can be easily volatilized.

Solvents such as acetone or ethanol may be used for this purpose. If a solvent such as chloroform is used, adequate safety measures must be taken. The amount of sample applied relates to its sensitivity of detection, as long it does not overload the system. Preferably 10 µl of the sample solution (0.1 to 1.0% w/v) is applied in a circular spot to represent 1–1000 µg of the sample. Forced hot air can be used for vaporization of the solvent, with due considerations for safety and stability of the sample.

If the solute of interest is present in low concentration in the provided sample, a number of necessary steps can be taken to concentrate the analyte:

- Extraction can be used to judiciously extract the analyte, which can then be concentrated before applying on the chromatography paper.
- Alternatively, a larger volume can be applied on the paper. In this case, it may be desirable to apply the solution repeatedly in small portions to keep the size of the spot small.
- Another way is to apply the sample in a row of adjacent spots or as a streak. The object is to apply adequate sample for detection and/or quantitation while at the same time assuring that the chromatographic system is not overloaded. Overloading frequently causes tailing.
- Ionic substances from aqueous biological materials can be concentrated, if desirable, by using ion-exchange papers.

III. SAMPLE CLEANUP

Samples that contain a large amount of extraneous material have to be cleaned up before application on a chromatographic paper. Included in this category are notable examples of samples used for monitoring pollution of soil or water, animal feed samples, and samples from biological matrices.

A. Sample Solubility

The solubility information on the analyte and other known interfering materials may help design a cleanup procedure for many samples. For example, if the analyte of interest is soluble in ethanol and the extraneous material is not, ethanol extraction may be all that is necessary to extract the sample. However, it would be necessary to assure that essentially all the analyte has been extracted by carrying out additional extraction(s) or preparing a simulated sample.

B. Removal of Inorganic Materials

Inorganic substances can be removed by utilizing a solvent that selectively dissolves components of interest, or the appropriate ion exchangers have to be used.

C. Removal of Proteins

For removal of proteins, the usual methods are based on precipitation with tungstic acid. Alternatively, heat coagulation or methanol with acetone can be used. A more gentle procedure entails ultrafiltration. In some cases, the proteins can be separated from the analyte by selective retention of low molecular weight ionic substances on ion exchangers, which can then be eluted with an appropriate solvent for paper chromatography.

D. Removal of Lipids

Lipids can be removed by extraction with lipophilic (hydrophobic or nonpolar) solvents. Alternatively, the paper chromatogram can be first developed with a lipophilic solvent to remove the lipids from the non-lipophilic analyte of interest. Since the analyte is not lipophilic, it should remain at the origin. The analyte can then be chromatographed with suitable solvents.

IV. DERIVATIZATION

At times, it is appropriate to form a derivative to chromatograph the desired analyte. Derivatization can achieve the following.

1. Improve the detectability of the analyte.
2. Assure that volatile analytes such as alcohols do not volatilize during chromatography or any of the other steps used to achieve separation.
3. Remove extraneous materials.
4. Improve the separation of materials.

Derivatization is generally carried out on the basis of the reactivity of various functional groups of the analyte.[2] This suggests that it is important to select suitable reaction conditions to obtain an appropriate derivative. Ideally, derivatization should be rapid and quantitative with minimum by-products; the excess reagents should not interfere, or they should be easily removed.

Most compounds are neutral, acidic, basic, or amphoteric. By derivatizing all of the functional groups, the detectability and chromatographic characteristics of a compound can be improved. Since UV is commonly used for detecting compounds separated by paper chromatography, the primary goal of a derivatization is to increase UV absorbance of the compounds of interest.

The derivatives are prepared by attaching a chromophore with high molar extinction coefficient. The reaction with a specific functional group not only provides the desired derivative, but also by lowering polarity, can offer a derivative with better chromatographic properties. Some of the common UV-absorbing derivatives are given in Table 1. Derivatives can be prepared for the compounds containing amino, carboxyl, hydroxy, or carbonyl groups. The compounds containing both amino and carboxyl groups can be derivatized with phenyl isothiocyanate to yield useful derivatives.

The derivatives that increase fluorescence are given in Table 2. Again, derivatization reagents have been provided for the amino, carboxyl, hydroxyl groups as well as amino acids.

These derivatization procedures should be employed only when they are absolutely necessary, because they add an additional burden in terms of time

TABLE I Common UV-Absorbing Derivatives

Compound	Reagent	Derivative
R$_2$NH	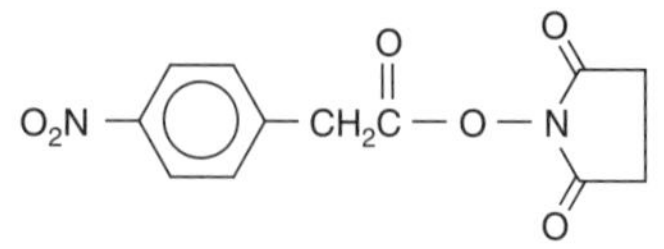 O$_2$N—⟨⟩—CH$_2$C(=O)—O—N(succinimide)	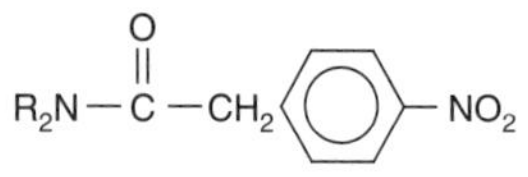 R$_2$N—C(=O)—CH$_2$—⟨⟩—NO$_2$
RCOOH	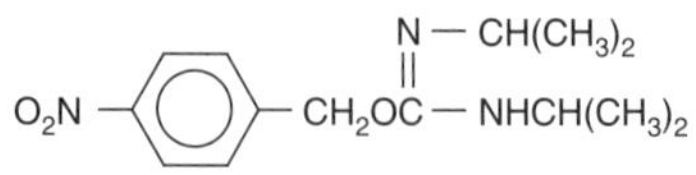 O$_2$N—⟨⟩—CH$_2$OC(=N—CH(CH$_3$)$_2$)—NHCH(CH$_3$)$_2$	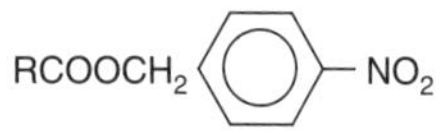 RCOOCH$_2$—⟨⟩—NO$_2$
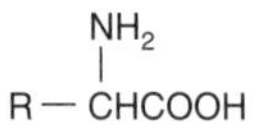 R—CH(NH$_2$)COOH	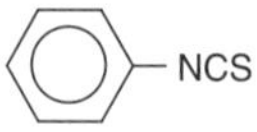 ⟨⟩—NCS	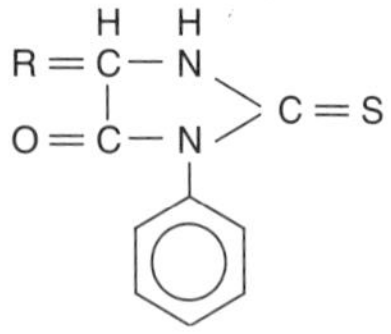 R=C(H)—N(H), O=C—N(H), >C=S, with phenyl
ROH	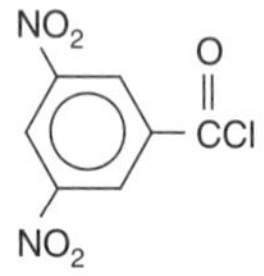 (NO$_2$)$_2$C$_6$H$_3$—C(=O)Cl	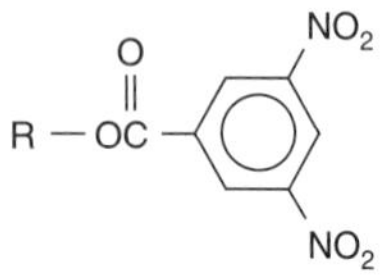 R—OC(=O)—C$_6$H$_3$(NO$_2$)$_2$
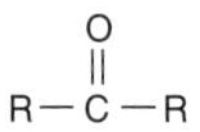 R—C(=O)—R	 O$_2$N—⟨⟩—CH$_2$ONH$_2$·HCl	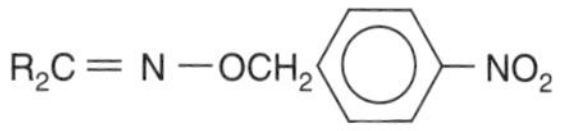 R$_2$C=N—OCH$_2$—⟨⟩—NO$_2$

and they require careful monitoring to assure that no material is lost during derivatization.

Derivatization reactions can also be used for better detection of the separated compound after the chromatogram has been developed (see Section VII).

V. MOBILE PHASES AND STATIONARY PHASES

A. Solvents

A list of common solvents that may be used in a mobile phase is given below in the order of increasing eluting power from alumina, known as eluotropic series of Trappe.

Petroleum ether
Carbon tetrachloride
Trichloroethylene
Isopropyl ether
Toluene
Benzene
Chloroform
Diethyl ether
Methyl ethyl ketone
Acetone

7 PAPER CHROMATOGRAPHY

TABLE 2 Common Fluorescent Derivatives

Compound	Reagent	Derivative
RNH₂	5-(dimethylamino)naphthalene-1-sulfonyl chloride (SO₂Cl)	dansyl sulfonamide (SO₂NHR)
	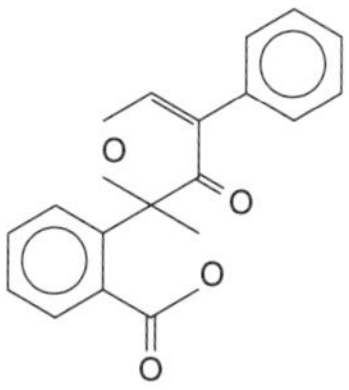 phenyl-substituted isobenzofuranone	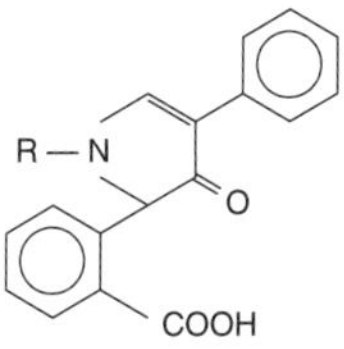 R–N substituted derivative with COOH
R–C(=O)–OH	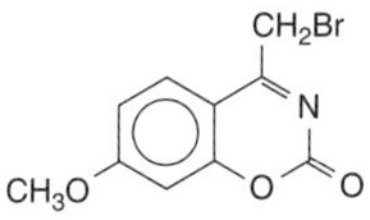 4-bromomethyl-7-methoxycoumarin (CH₂Br)	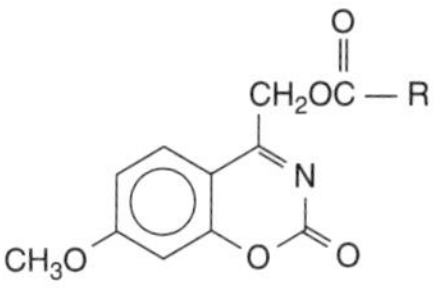 4-(acyloxymethyl)-7-methoxycoumarin (CH₂OC(=O)–R)
R–CH(NH₂)COOH	pyridoxal (HO, C(=O)H, CH₂OH, H₃C, N)	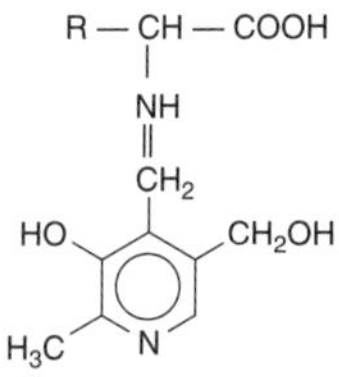 Schiff base derivative (R–CH–COOH, NH=CH₂, HO, CH₂OH, H₃C, N)
Ar–OH	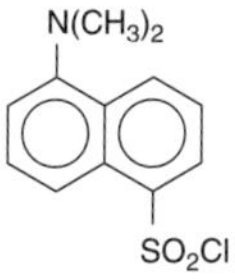 dansyl chloride (SO₂Cl, N(CH₃)₂)	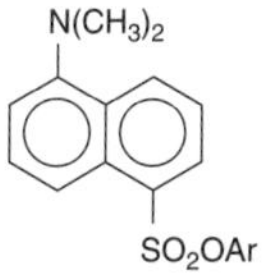 dansyl ester (SO₂OAr, N(CH₃)₂)

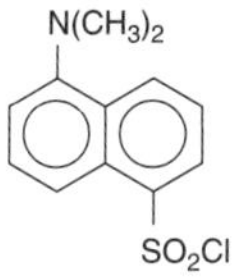

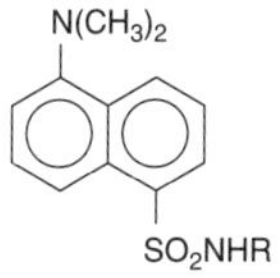

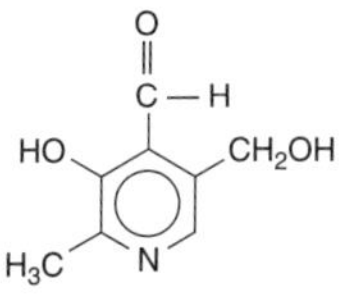

Ethyl acetate
Dioxane
Acetonitrile
Ethanol
Methanol
Water

A series of solvents similar to eluotropic series for adsorption chromatography can be thus prepared especially for paper chromatography (see Table 3). The solvents can be used singly or in a mixture in the mobile phases. They have been classified on the basis of their ability to form hydrogen bonds. At the head of this series are solvents that are either a donor or an acceptor of electron pairs and have the ability to form intermolecular hydrogen bonds (hydrophilic or polar solvents), whereas the solvents at the end of the series lack this property (hydrophobic, lipophilic, or nonpolar solvents).

TABLE 3 Useful Solvents for Paper Chromatography

Miscible with water (in all proportions)	Not fully miscible (with water)
Formamide	t-Butanol
Methanol	n-Butanol
Acetic acid	n-Amyl alcohol
Ethanol	Ethyl acetate
Isopropanol	Diethyl ether
Acetone	Toluene
n-Propanol	Petroleum ether

By reviewing Table 3, it can also be determined whether a given solvent is miscible with water. This is an important consideration when a mobile phase is being prepared for paper chromatography (see Section V. C). Starting with formamide up to tertiary butanol, the solvents are miscible with water in all proportions, whereas the rest of the solvents are not fully miscible.

Appropriate selection of a single solvent or a mixture of hydrophilic solvents can influence the mobility of ionizable substances. In addition, variation of pH can significantly improve the separation because ionization of a solute, based on its pK_a value, can be influenced by the pH.

B. Stationary Phases

As mentioned before, the solvent(s) held on paper is called the stationary phase in paper chromatography. Stationary phases may be divided into two main categories: polar stationary phases and nonpolar stationary phases.

1. Polar Stationary Phases

The stationary phase held on the paper is either water or some other polar solvent, and the immiscible solvents pass through as the mobile phase. This form of chromatography may be called normal-phase chromatography. The paper is generally not impregnated with the aqueous phase, although this may be useful in some cases. The paper takes up moisture when it is suspended in the closed chamber where the atmosphere is saturated with water vapor. When an aqueous buffer solution or a salt solution is used as the stationary phase, the paper is drawn through the solution and dried prior to development. Other polar stationary phases can be alcohol, formamide, propylene glycol, and so forth.

2. Nonpolar Stationary Phase

A hydrophobic solvent is held as the stationary phase and the hydrophilic mobile phase passes through. Since the migration of zones is reversed, this technique is called reversed-phase paper chromatography. If paraffin oil or silicone oil is used as the stationary phase, the paper is impregnated with the oil by using petroleum ether as the diluent. Alternatively, the hydrophilic nature of the paper can be modified chemically by esterification. The paper will then accept hydrophobic solvents from the atmosphere of the chamber.

C. Mobile Phases

A large number of mobile phases have been used; however, this number can be reduced significantly. Some of the useful mobile phases are listed below on the basis of their suitability for the materials that can be roughly classified as follows:

- Hydrophilic
- Hydrophobic
- Moderately polar

1. Hydrophilic Substances

The commonly used mobile phases include:

- *n*-Butanol/acetic acid/water (4/1/5)
- Isopropanol/ammonia/water (9/1/2)

The ratio of solvents is made on a volumetric basis. The mixture with butanol yields a two-phase system. All the components are shaken together, and the mixture is used to saturate the tank; however, the less hydrophilic system is used for development.

2. Hydrophobic Substances

Some of the useful mobile phases are listed here.

- Dimethyl formamide/cyclohexane
- Solution of 50% DMF in ethanol used for impregnation
- Paraffin oil/dimethyl formamide/methanol/water (10/10/1)
- Solution of 10% paraffin oil in an aromatic hydrocarbon used for impregnation

3. Moderately Polar Substances

For moderately polar substances, formamide with a number of solvents has been used. In all formamide systems, 40% ethanolic formamide solution is used for impregnation of the paper. The formamide may contain, depending on the substance to be analyzed, 5% ammonium formate or 0.5% formic acid for acidification of the stationary phase.

The simplest system is composed of a mixture of formamide and an aromatic hydrocarbon such as benzene or chloroform:

- Formamide/chloroform
- Formamide/toluene

Varying proportions of cyclohexane may be added to aromatic hydrocarbon mixtures to improve separations.

VI. DEVELOPMENT OF CHROMATOGRAMS

The following four modes of paper chromatography can be performed, depending on the flow of the mobile phase:

- Ascending paper chromatography: mobile phase flows upward.
- Descending paper chromatography: mobile phase flows downward.
- Horizontal paper chromatography: mobile phase flows horizontally.
- Radial paper chromatography: mobile phase flows radially.

Ascending or descending chromatography may be accomplished in a chamber as shown in Figure 2. Descending chromatography is most commonly used because chromatograms can be developed to longer distances, unhampered by gravity. It is well known that gravity affects solvent advancement by capillary action of paper in the ascending mode of paper chromatography. The longer distance development allows for better resolution because of a corresponding increase in the number of theoretical plates. Recall the concept of theoretical plates discussed earlier for distillation and column chromatography. The length of paper corresponds to the length of distillation or chromatographic column.

The chamber design for horizontal chromatography is not difficult because all one has to assure is that solvent is fed horizontally to the paper, which is also kept in a horizontal position.

Radial chromatography may be carried out in a closed petri dish. In radial paper chromatography, the sample is spotted in the center of a circular paper and the solvent is applied in the center by a wicking technique so that development of the chromatogram is toward the periphery.

Paper chromatograms are developed to a predetermined distance that allows calculation of the R_f value of each component after development:

$$R_f = \frac{\text{Distance traveled by the sample component}}{\text{Distance traveled by the developing solvent}}$$

In case R_f values are too low, it may be desirable to carry out overrun development, where a descending technique is frequently used and development is not stopped when the solvent front reaches the lower end of the paper. As a matter of fact, it is allowed to overrun for the desired time.

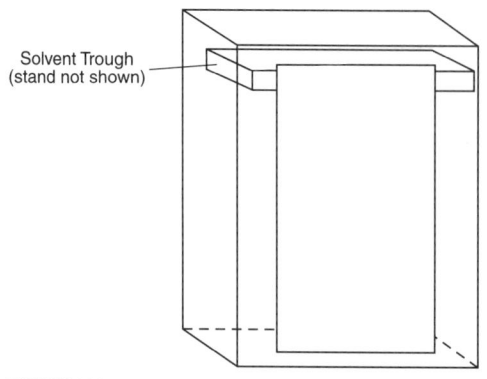

FIGURE 2 Developing tank for descending paper chromatography.

Two-directional development may be carried out with the same phases or two different phases to enhance separations. In this case, it is preferable to use a square-shaped paper. The paper is turned 90 degrees after the first development, and the second development is carried out with the same or different mobile phase.

VII. DETECTION

The chromatogram is dried at room temperature, or higher temperature may be used, such as forced hot air, depending on the solvents that were employed and the stability of the resolved samples. In paper chromatography, the separated components are identified generally on the basis of their mobility in the mobile phase. This mobility is expressed by R_f value as defined previously.

Sometimes the mobility values are expressed as R_f values relative to a known reference material. The R_f values are generally constant under strictly controlled conditions; however, duplication of these conditions from one day to the next or over a longer period is difficult to achieve for paper chromatographic conditions. Temperature and humidity play a very significant role and need to be rigorously controlled.

The colored materials can be detected visually; however, for colorless materials, it is necessary to use alternate methods. UV light between 250 and 260 nm can be used for a large number of compounds. Many organic substances fluoresce after irradiation with UV light from a high-pressure mercury vapor lamp. Selective detection reagents that owe their selectivity to reaction with certain functional groups or to the presence of unsaturation may also be used (see Table 4).

Enzymatic and microbiological methods have been found to be useful. For example, enzymatic methods can be put to use for detection of enzymes rather than the substrate. Amylase may be detected by spraying the paper with starch solution, incubating for a suitable period, then spraying with iodine vapors. Amylases appear as white spots on a blue background. Samples such as pesticides that inhibit enzymes can be detected by appropriate adjustment of the method.

TABLE 4 Selective Spray Reagents

Reagent	Observation	Applications
Iodine vapors	Brown spots	Unsaturated organic compounds
2′,7′-Dichlorofluorescein	Yellow-green spots under UV light at 254 nm	Most organic compounds
0.1–0.5% Bromcresol green	Yellow spots	Carboxylic acids
50% SbCl$_3$ in HOAc	Various colors	Steroids, alicyclic vitamins, carotenoids
0.3% Ninhydrin in n-butanol containing 3% HOAc	Pink/purple spots	Amino acids

In microbiological methods, the inhibition of microbial growth is monitored. These methods are also called bioautographic methods. The chromatogram is placed on a large petri dish with the appropriate microorganism in a nutrient medium. On incubation at a suitable temperature, the microorganism grows all over the petri dish except where the sample has been placed, inhibiting growth.

VIII. QUANTITATION

Semiquantitative estimations can be made against standards run comparably. It is generally preferable to bracket the sample concentration with suitable standards.

Quantitation requires eluting the spot of interest and then analyzing the solution by colorimetric, UV, or other suitable detection techniques. Standards are similarly run and treated. The elution of substances must be done quantitatively. The location of the spot of interest is accomplished by employing the detection techniques discussed earlier. A surrogate detection technique may also be used in case the quantitation technique is likely to be different from the detection technique. The sample should be eluted in a minimum volume. Automatic elution devices are also available. The eluted spot then can be analyzed by the technique of choice for a given sample.

Densitometric techniques for visible–UV wavelengths can be used in any of the following modes:

- Transmittance
- Reflectance
- Fluorescence

The paper strip is run automatically over a slit positioned in front of a photocell. Either the light reflected from the paper or that transmitted through the paper is measured. It is necessary to run standards and scan them in a similar manner. The above discussion relates to visible–UV determinations.

The same principles are applied to determination of substances that provide UV-excited fluorescence. Radioscanners can be used for monitoring radioactive samples.

REFERENCES

1. Macek, K. *Chromatography* (E. Heftman, Ed.), Van Nostrand, Reinhold, 1975.
2. Ahuja, S. *Ultratrace Analysis of Pharmaceuticals and Other Compounds of Interest*, Wiley, NY, 1986.

QUESTIONS FOR REVIEW

1. Define paper chromatography.
2. Describe various modes of chromatography.
3. What are the advantages offered by paper chromatography?
4. Name some universal detection reagents for paper chromatography.

8
THIN-LAYER CHROMATOGRAPHY

I. STATIONARY PHASES FOR TLC
   A. Adsorption
   B. Partition
   C. Ion Exchange
   D. Size Exclusion
   E. Miscellaneous
II. TLC OF ENANTIOMERIC COMPOUNDS
   A. Achiral Stationary Phase and Achiral Mobile Phase
   B. Achiral Stationary Phase and Chiral Stationary Phase Additives
   C. Chiral Stationary Phase and Achiral Mobile Phase
III. SAMPLE APPLICATION
IV. MOBILE PHASES
   A. Commonly Used Mobile Phases
V. DEVELOPMENT OF CHROMATOGRAMS
   A. Effect of Temperature
   B. Gradient Elution
   C. Miscellaneous Techniques
VI. DETECTION AND QUANTITATION
VII. APPLICATIONS
   A. Amino Acids
   B. Pharmaceuticals
   C. Vitamins
   D. Dyes
   REFERENCES
   QUESTIONS FOR REVIEW

Thin-layer chromatography (TLC) is an open-bed chromatographic technique that is generally carried out on a thin layer of stationary phase coated on a glass plate. It is routinely used in many laboratories in the chemical/pharmaceutical and related industries, for both qualitative and semiquantitative work. Quantitative analyses, of course, can also be performed. Some laboratories find this technique extremely useful and claim that very good precision can be achieved, even at very low levels of analyses.[1] However, elaborate steps must be taken to assure precision. These are discussed at some length later in this chapter

    The origin of TLC can be traced back to 1938 when two Russian researchers, Izmailov and Shraiber, utilized a technique called drop chromatography on horizontal thin layers. It took another twenty years for this technique to become a practical tool when Stahl described equipment and efficient sorbents for the preparation of plates (for more details, see Stahl's book listed in Reference 2). The technique suffered from relatively low efficiency for another twenty years before making a great advance that resulted from introduction of smaller and more uniform particle sizes in the sorbents. Greater reproducibility in the preparation of layers has helped to achieve better reproducibility of results. These developments led to the introduction of TLC plates that offer high resolution because of the smaller particle size of the sorbent used in their preparation and other special steps taken to ensure a more uniform compact layer. Thus, the term high-performance thin-layer chromatography (HPTLC) was brought about. Unfortunately, it is not uncommon today to use the terms TLC and HPTLC interchangeably. This should be avoided because it is more appropriate to use the term HPTLC when it can be clearly demonstrated that the given plate offers higher resolution based on the above-mentioned considerations; that is, a higher number of theoretical plates are being utilized than offered by an average TLC plate.

The main reason for the wide use of the technique of thin-layer chromatography is the quality of information provided by it in a very short time. Furthermore, it is relatively easy to learn and can be performed inexpensively. Although TLC is still being used primarily as an adsorption technique, it has been found useful for separations based on ion exchange, partition, and other modes used in chromatography. Another significant advantage of this technique is that two complementary mobile phases can be employed consecutively, in a two-directional development, to provide additional information.

TLC can also be used as a preparative technique for isolation and purification of compounds of interest. The layer of sorbent used for this purpose is generally thicker than a normal TLC plate. However, normal plates can be used when only a small quantity of material is needed or when a large number of repetitive runs have to be made. An important point to remember here is that TLC is commonly used for preparative separations because the method development and operation requires minimal time and cost investment.

I. STATIONARY PHASES FOR TLC

A number of different stationary phases can be used for TLC that permit performance of a variety of modes of chromatography. The coating material is generally applied as a thin coating (e.g., 250-μm thick coating of silica gel) in a suitable solvent on a glass plate (20 × 20 cm). Other plate sizes, as well as plates of other materials, have also been used. Most of the commonly used TLC plates are available commercially, so it is not necessary to prepare them in the laboratory.

As mentioned before, in all chromatographic methods including TLC it is necessary to have two phases to achieve a successful separation. These phases are designated as the stationary phase and the mobile phase. In conventional

TLC, the stationary phase is generally silica gel and the mobile phase is composed of a mixture of solvents. Contrary to common belief, the observed separation in TLC is not due to adsorption on silica gel alone. There is always a finite amount of water present in the silica gel plates that acts as a partitioning agent. Furthermore, the mobile phase solvents are adsorbed onto silica during development and thus provide yet another mechanism for partition. Ionic sites in silica gel permit ion exchange, and metallic impurities provide mixed mechanisms for certain separations. The points made here regarding the presence of other components on the mechanism of a given separation should be given due consideration prior to designing a separation by TLC.

Various materials have been used for coating thin-layer plates to offer the following modes of chromatography:[2]

- Adsorption
- Partition
- Ion exchange
- Size exclusion
- Miscellaneous

A. Adsorption

Some of the adsorbents that have been used are as follows:

- Silica gel (silicic acid)
- Alumina (aluminum oxide)
- Kieselguhr (diatomaceous earth)

Of these adsorbents, silica gel is most commonly used in TLC because of its versatility.

B. Partition

The following phases can be used for performance of partition TLC:

- Cellulose
- Kieselguhr (diatomaceous earth)
- Silica gel

It should be noted here that normal- and reversed-phase TLC discussed in Section I. E can also be considered partition chromatography.

C. Ion Exchange

The following materials have been used for ion-exchange TLC:

- Cellulose phosphate
- Polyethylenimine cellulose
- DEAE cellulose

Both cation and anion exchange can be performed by TLC, depending on the ion exchanger coated on the plate.

D. Size Exclusion

The following dextran gels have been utilized:

- Sephadex G-25
- Sephadex G-50
- Sephadex G-75
- Sephadex G-100

It is important that the dextran gels used for size-exclusion TLC are superfine.

E. Miscellaneous

A variety of other modes of chromatography have been used in TLC that do not necessarily fit the above classifications.

I. TLC on Polyamides

A number of polyamides mentioned here have been used in TLC to achieve some interesting separations of tannins, coumarins, and flavone glycosides:

- *e*-Polycaprolactam
- Acetylated *e*-polycaprolactam
- Polyacrylonitrile

The exact mechanism of separation in these cases has not been fully defined. However, it appears that the separation results in groups of compounds that have the same number of hydrophilic functional groups.

2. Normal-Phase TLC

The majority of TLC separations are performed on silica gel with a relatively nonpolar mobile phase. This type of separation is called normal-phase separation to distinguish it from reversed-phase separation discussed next. Both adsorption and partition play a role in these separations. It is important to point out that partition-based separations are possible by minimizing the adsorbent activity of silica and enhancing the potential for partition by selection of suitable stationary phases similar to those used for normal-phase HPLC (see Chapter 10).

3. Reversed-Phase TLC

In reversed-phase separation, the stationary phase is nonpolar and the mobile phase is relatively polar. In the early stages of development of this technique, reversed-phase TLC was performed by utilizing paraffin oil on silicic acid or kieselguhr as the stationary phase. Acetylated cellulose has also been used. It is important to remember that this allows partition-based separations.

More recently, it has become possible to use plates made with various chain-length alkanes (C-2 to C-18) that are chemically bonded to silica to produce a stationary phase that is significantly nonpolar. We will see in Chapter 10 that these phases are also commonly used in high-pressure liquid chromatography (HPLC), a technique that offers high resolution in a short time. These TLC plates can be used to develop excellent separations or can help scout or optimize a mobile phase for HPLC. Surface coverage of these alkane stationary phases can vary, and the unreacted silanol sites influence separations.

II. TLC OF ENANTIOMERIC COMPOUNDS

A variety of chromatographic approaches have been used to achieve the separations of enantiomeric compounds. The most important point considered in these separations is the chirality of the sample molecules. Therefore, TLC separation of these compounds is being covered here as a special group (a more detailed discussion is given in Chapter 11).

Various approaches that have been used to achieve separation of chiral compounds by TLC can be classified as follows:

- Achiral stationary phase and achiral mobile phase
- Achiral stationary phase and chiral stationary phase additive
- Chiral stationary phase (CSP) and achiral mobile phase

A significant amount of coverage is provided here to enable the reader to try TLC for separation of enantiomeric compounds. Since TLC is a very useful qualitative technique that entails minimal costs, it can be used as an independent technique with its well-known limitations of resolution and low precision. It can also provide good indications as to which HPLC method would be best suited for resolving enantiomers.

A. Achiral Stationary Phase and Achiral Mobile Phase

To utilize an achiral stationary phase and an achiral mobile phase, derivatization is required to separate the diastereomers. (−)-Menthyl chloroformate or (S)-camphor-10-sulfonyl derivatives are commonly prepared for this purpose. Because of the difficulty of preparation of derivatives and the associated problems of by-products, this technique does not have widespread use.

B. Achiral Stationary Phase and Chiral Stationary Phase Additives

1. β-Cyclodextrin and Other Additives

A number of chiral eluent additives have been investigated. Some of these are listed below:

- β-Cyclodextrin
- (−)-Brucine
- D-Galacturonic acid
- (+)-Tartaric acid

β-Cyclodextrin is a chiral toroidal molecule with a finite cavity formed by the connections of seven glucose units via 1,4-linkage (see Fig. 1). Retention is influenced by the size of the cavity of the cyclodextrin and other associated interactions of the enantiomers with the oligomer.

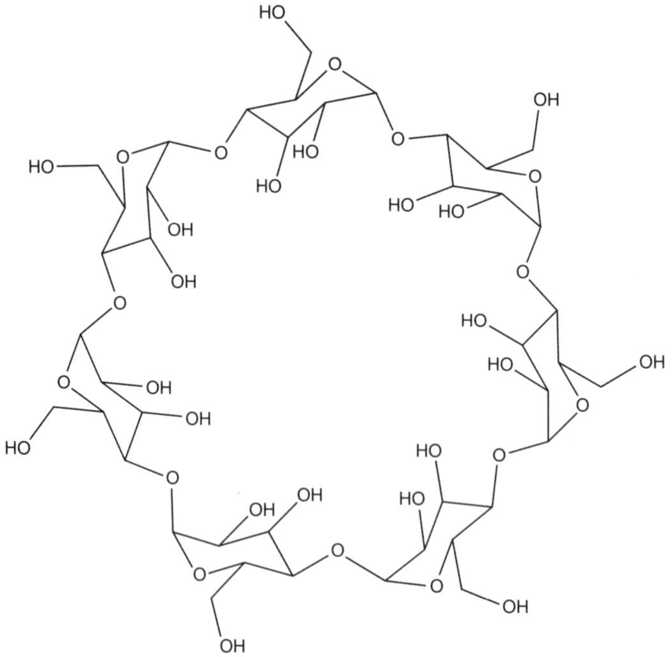

FIGURE 1 Structure of β-cyclodextrin.

C. Chiral Stationary Phase and Achiral Mobile Phase

The TLC plates can be coated with the following materials to provide useful stationary phases for enantiomeric separations:

- Cellulose
- Chiral compounds
- Cyclodextrin
- Ligands

1. Cellulose Plates

Cellulose is a linear macromolecule composed of optically active D-glucose units with helical cavities (see Fig. 2). The separation of enantiomers is influenced by their different fit in the lamellar chiral layer structure of the support.

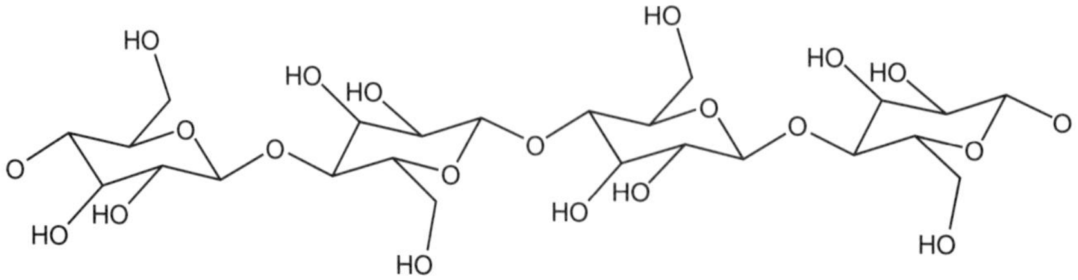

FIGURE 2 Structure of cellulose.

Peracetylation of the cellulose can be performed in a way to assure that conformation and relative position of the carbohydrate bands in their crystalline domain remain intact. Systematic investigations of this chiral support has resulted in commercialization of a microcrystalline triacetylcellulose plate. These plates are stable with aqueous eluent systems and are resistant to dilute acids and bases. They are also stable in alcoholic eluents but are attacked by glacial acetic acid and ketonic solvents.

2. Chiral Compounds on a Plate

A notable example of such plates is silica gel/(+)-tartaric acid plates. Plates with brush- or Pirkle-type phases contain a selected compound, for example, (R) or (S) isomer of N-(3,5-dinitrobenzyl) phenylglycine.

3. β-Cyclodextrin Plates

Various cyclodextrins (i.e., α-, β-, or γ-) have been used. Out of these cyclodextrins, β-cyclodextrin is the one most commonly used. It may be recalled from Figure 1 that β-cyclodextrin is a cyclic oligosaccharide with a doughnut-shaped structure that is composed of seven α-L-glucose units linked through the 1,4 position.

4. Ligand-Exchange Plates

Merck RP-18 WF 254 S plates covered with copper acetate and LNDH (L-N-*n*-decylhistidine) can serve as a good example of this type of plate.

5. Commercial TLC Plates

Chiralplate was the first ready-to-use TLC. It was marketed by Macherey-Nagel. This was followed by the chiral HPTLC plate, CHIR, in 1988, which further simplifies the use of TLC for chiral separations. In 1993, Astec offered β-cyclodextrin plates for evaluating chiral separations prior to HPLC.

III. SAMPLE APPLICATION

The sample is generally dissolved in a solvent that can be easily volatilized. Ten- to 100-µl volume of the solution is applied for analytical TLC at a few centimeters above the base of the plate. An attempt is made to apply a circular spot with small diameter. This requires that a larger volume be applied, when necessary, in repetitive small portions. The weight of the sample deposited at the origin relates to detectability of the sample and its potential for tailing. Samples susceptible to oxidation can be spotted under a nitrogen atmosphere.

Sample preparation and application achieve even greater importance when trace or ultratrace analysis has to be performed. The most important requirement for excellent performance of TLC, that is, optimal sample application, is generally overlooked as a factor in the ability to detect ultratrace components or obtain separation.[1] The average practitioner applies too much sample and overloads the system; consequently he or she wonders why no separation was achieved. It is important to apply as little sample as is consistent with the ability to detect the analyte of interest. This requires proper sample preparation as well as sample application. So sample preparation can be viewed as a problem of concentration of analyte.

The concentration of sample can require transfer of desired components in a sample from the original sample matrix to a smaller volume of a selected solvent or a mixture of solvents. The technique utilized for a given sample is determined by the relative volatility of the desired components and the matrix and the chemical nature of components of interest and matrix.

For liquid and solid samples, the obvious familiar separation techniques that can lead to a concentrated sample such as extraction, evaporation, distillation, and precipitation can be used. However, in trace or ultratrace analysis, extraction and/or evaporation may be the only two techniques that help minimize severe losses of analyte.

In biological samples where organic components are sought at low levels, short columns that retain components of interest can be used. A detailed discussion on these topics may be found in Chapter 5 of Reference 3. Listed here are several approaches that can be used for sample preparation:[3]

- Precipitation of undesirable components
- Liquid–solid extraction
- Liquid–liquid extraction
- Ultrafiltration
- Ion-pair extraction
- Derivatization
- Complex formation
- Freeze-drying
- Solid-phase extraction columns
- Ion exchangers
- Specialized columns
- Preconcentration
- Precolumn switching
- Cleanup procedures
- Miscellaneous

Some of the column extraction procedures shown in Figure 3 are for organic substances with molecular weights greater than 2000. The samples are further divided on the basis of whether they are soluble in water or organic solvents. A suitable extraction column can then be selected to perform sample preparation, depending on the type of sample and contaminants. For example, a reversed-phase chromatography column is used for samples that are soluble in organic solvents. Ion-exchange chromatography, size-exclusion chromatography, and reversed-phase chromatography can be used for water-soluble materials.

Figure 4 shows sample preparation for organic substances with molecular weights less than 2000, based on the solubility of the sample; that is, whether the sample is soluble in water or an organic solvent. The choice of columns in this case is broad and includes liquid–solid chromatography and normal-phase chromatography, along with reversed-phase chromatography. Ion-exchange chromatography in addition to various chromatographic techniques discussed earlier can be used for water-soluble materials.

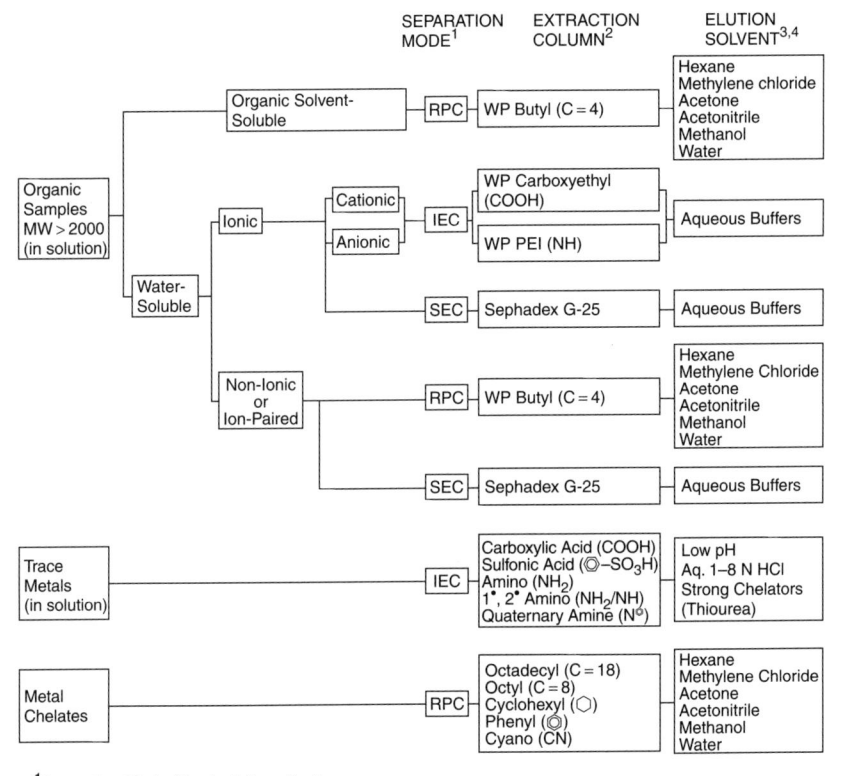

8 THIN-LAYER CHROMATOGRAPHY

FIGURE 3 Sample preparation for various compounds.

			SEPARATION MODE[1]	EXTRACTION COLUMN[2]	ELUTION SOLVENT[3,4]

Organic Samples MW > 2000 (in solution)
- Organic Solvent-Soluble → RPC → WP Butyl (C = 4) → Hexane, Methylene chloride, Acetone, Acetonitrile, Methanol, Water
- Water-Soluble
  - Ionic
    - Cationic → IEC → WP Carboxyethyl (COOH) → Aqueous Buffers
    - Anionic → IEC → WP PEI (NH) → Aqueous Buffers
  - → SEC → Sephadex G-25 → Aqueous Buffers
  - Non-Ionic or Ion-Paired → RPC → WP Butyl (C = 4) → Hexane, Methylene Chloride, Acetone, Acetonitrile, Methanol, Water
  - → SEC → Sephadex G-25 → Aqueous Buffers

Trace Metals (in solution) → IEC → Carboxylic Acid (COOH), Sulfonic Acid (−SO$_3$H), Amino (NH$_2$), 1°, 2° Amino (NH$_2$/NH), Quaternary Amine (N°) → Low pH Aq. 1–8 N HCl, Strong Chelators (Thiourea)

Metal Chelates → RPC → Octadecyl (C = 18), Octyl (C = 8), Cyclohexyl, Phenyl, Cyano (CN) → Hexane, Methylene Chloride, Acetone, Acetonitrile, Methanol, Water

[1] Separation Mode (Vendor information)
LSC: Liquid-Solid Chromatography (Absorption)
NPC: Normal-Phase Chromatography (Bonded-Phase Partition)
RPC: Reversed-Phase Chromatography (Bonded-Phase Partition)
IEC: Ion-Exchange Chromatography (Bonded-Phase Ion Exchange)
SEC: Size-Exclusion Chromatography

[2] Extraction columns listed in degree of increasing polarity
[3] Eluting solvents listed in degree of increasing polarity
[4] Selective elution can be performed by combining two or more miscible solvents to achieve various degrees of polarity

The advent of sample preparation columns or cartridges has overcome many sample preparation problems. However, the operator must evaluate the use of a given method for a particular situation. A typical example is the use of a Sep-pak (Waters Associates) cartridge for removal of cortisone from urine for chromatographic analysis. The example given here highlights analysis of 50 ng of analyte in as much as 2 liters of matrix.

A small volume of urine is passed through a Sep-pak C-18 cartridge that has been pretreated with a small volume of methanol, followed by a small volume of water. The interfering materials are removed by passing methanol/water (20:80) through the cartridge. Cortisol is then eluted with as little as 2 mL of methanol. The sample is concentrated by evaporating the methanol, or an aliquot is analyzed by HPTLC. It is recommended that a known amount of radioactive material be added to assess the recovery. (Similar cartridges are also available from other suppliers.)

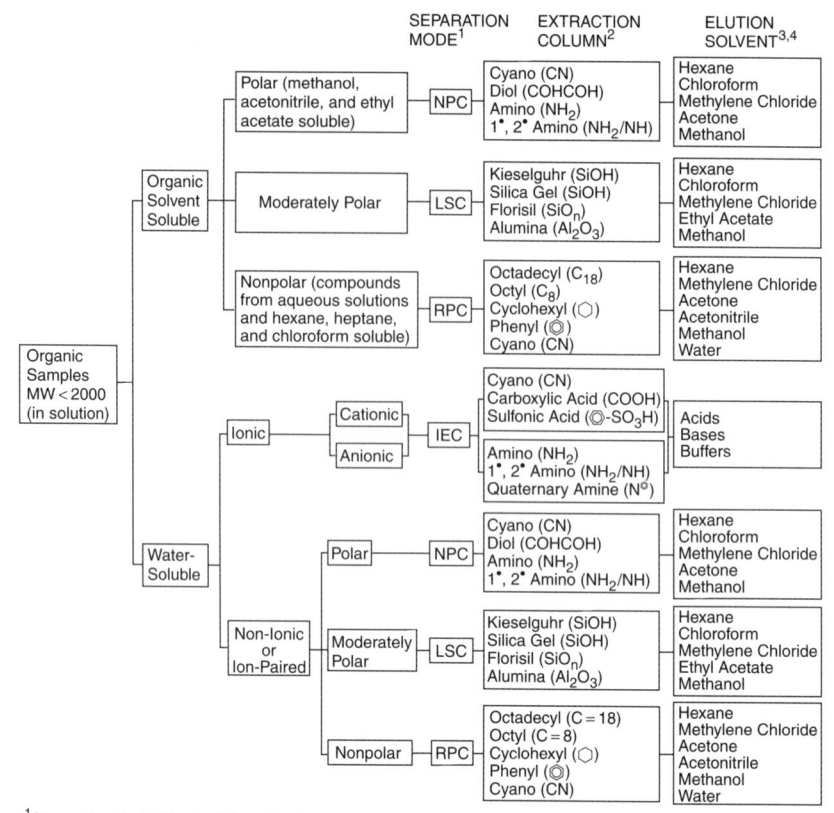

FIGURE 4 Sample preparation based on solubility considerations.

TLC may also be used for preparing a suitable clean sample for analysis by other techniques that are more sensitive and, as a result, are also more sensitive to interfering materials. The crude samples are spotted on the plate, and plates are developed with a suitable mobile phase that would leave the interfering materials at the origin but move the desired component(s) to the middle of the plate. The desired component is then eluted and analyzed by more sensitive techniques.

For preparative separations, thicker plates (>250-μm thick coating) are used to allow application of higher weights of sample, and the sample is frequently applied as a streak. If normal thickness plates are used, a much larger number of plates must be developed to obtain the desired quantity of material.

TABLE 1 Polarity of Solvents Based on Rohrschneider Data

Solvent	Polarity
n-Hexane	0.0
Toluene	2.3
Diethyl ether	2.9
t-Butanol	3.9
Tetrahydrofuran	4.2
Ethyl acetate	4.3
Chloroform	4.4
Dioxane	4.8
Ethanol	5.2
Acetone	5.4
Acetic acid	6.2
Acetonitrile	6.2
Methanol	6.6
Formamide	7.3
Water	9.0

IV. MOBILE PHASES

The selection of a mobile phase depends on the mode of chromatography that is being used for a given sample. In adsorption chromatography, a single solvent from the eluotropic series (see Chapter 7) can be used. However, it is more common to use a mixture of solvents rather than a single solvent. Acidic and basic components help to further improve separations. To assure a better reproducibility, it is desirable to use a simple mixture. Table 1 provides information on polarity of solvents based on Rohrschneider data.

Some of the systems used for paper chromatography may be also used for TLC on cellulose layers. As a matter of fact, the separations achieved on cellulose layers are generally better than those observed on paper. The mixtures of solvents that have been commonly employed as mobile phases for various modes of chromatography are given in the next section.

A. Commonly Used Mobile Phases

Some of the common mobile phases utilized in various modes of thin-layer chromatography are listed here.

1. Adsorption TLC
   - *n*-Hexane/diethyl ether (1:1)
   - *n*-Hexane/diethyl ether/acetic acid (90:10:1)
   - Chloroform/methanol (19:1)
   - Toluene/acetone (1:1)

2. Partition TLC
   - *n*-Butanol/acetic acid/water (4:1:5)
   - *n*-Butanol/acetone/diethylamine/water (10:10:2:5)
   - 2-Propanol/formic acid/water (2:1:5)
3. Ion-Exchange TLC
   - 0.001–1 N hydrochloric acid
   - 0.001–1 N sodium hydroxide
   - 0.001–1 N sodium chloride
4. Reversed-Phase TLC
   - Acetonitrile/acetic acid (1:1)
   - Acetone/water (3:2)
   - Chloroform/methanol/water (15:5:1)
5. TLC on Polyamides
   - Ethanol/water (1:1)
   - Acetone/water (1:1)
   - Water/ethanol/acetic acid/dimethylformamide (6:4:2:1)
6. TLC on Dextran Gels
   - 0.01–0.1 N acetic acid
   - 0.01–0.1 N ammonia
   - 0.01–0.1 M sodium phosphate buffer

V. DEVELOPMENT OF CHROMATOGRAMS

Thin-layer chromatography can be performed in ascending, descending, or horizontal modes, as described in Chapter 7 for paper chromatography. The most commonly used mode in TLC, however, is the ascending mode; whereas, in paper chromatography the descending mode is generally favored. Recall that the solvent is allowed to travel up in the ascending mode. The sample is developed to a distance of 10 to 15 cm; that is, the distance of movement of the solvent front from the point where the sample is spotted. There are no special advantages to the descending technique in TLC; however, horizontal chromatography can help improve study of the theoretical aspects of chromatography.

The plate is developed in a closed chamber, which is generally lined with filter paper to saturate the chamber with the mobile phase. Saturation assures reproducibility of development conditions and allows some of the volatile solvents to be adsorbed onto the sorbent. Lining with filter paper is not necessary with specialized chambers such as the S-chamber or the BN-chamber. The R_f value is calculated for each separated component, as discussed for paper chromatography in Chapter 7. The R_f values provide a preliminary identification of a compound when compared with an authentic material run similarly (and preferably at the same time).

At times, it is preferable to use an unsaturated chamber because it saves time and sometimes produces resolution unattainable in a saturated chamber. This is due to the fact that unsaturation engenders longer development time that can be equated in some ways with using longer development length, and the effect of the adsorbed solvent from a saturated chamber is less dominant. However, reproducibility of the R_f values is generally a problem in this case.

A. Effect of Temperature

TLC is commonly performed at ambient temperature; however, at times separations can be improved by utilizing higher or lower temperatures. Therefore, the effect of temperature on separation should be explored.

B. Gradient Elution

Compounds that differ greatly in migration rate can be resolved by gradient elution, where the sample is developed from the center of the plate and the composition of applied mobile phase is changed continuously. Alternatively, gradient layers can be used.

C. Miscellaneous Techniques

Separations may also be improved by continuous development, in which the solvent is allowed to run over the top of the plate by evaporation, or it can be removed by selective techniques. Once again, this is equivalent to using a much longer plate length.

Of the various modes of development of a TLC plate, two-directional development with the same mobile phase or different mobile phases offers the greatest potential for obtaining better resolution in TLC.

Multiple development with the same mobile phase or with a different mobile phase will generally yield better separation than a single-pass development. Utilizing a different mobile phase can provide an added advantage of improving selectivity offered by the chosen mobile phase.

VI. DETECTION AND QUANTITATION

Colored materials are generally detected by simple visualization. Radioactively labeled compounds may be detected on thin-layer chromatograms by autoradiography or by special scanners.

Detection with UV is commonly employed for UV-absorbing compounds; both long- and short-wavelength UV light can be used for this purpose. Fluorescent indicators are incorporated in the adsorbent layer to assist in UV detection. The plates thus prepared for silica gel are called silica gel GF, and the sample is detected by fluorescent quenching against the background of a fluorescent plate under UV light.

Various spray reagents may be used to detect samples with ease on a TLC plate as opposed to chromatographic paper. Even charring with corrosive reagents such as sulfuric acid can be used, or iodine vapors can be used as

TABLE 2 Commonly Used Detection Methods

Treatment	Visualization	Applications
Visible light	Colors	Colored compounds
UV (254 or 366 nm)	Fluorescence	Many organic compounds
Dichlorofluorescein	Yellow-green spots under UV at 254 nm	Most organic compounds
Iodine vapors	Brown spots	Unsaturated organic compounds
Chromic sulfuric acid	Black spots on heating	Nonvolatile organic compounds
Bromocresol green	Yellow spots	Carboxylic acids
Ninhydrin	Pink/purple spots	Amino acids, amines
Antimony chloride	Various colors	Steroids, carotenoids

a universal detecting reagent. A list of the most commonly used detection reagents is given in Table 2.

Quantitation in TLC is possible by comparing the sample spot with the standard spot. The size of the spot increases with the amount of material it contains; however, this relationship is not necessarily linear. In adsorption TLC, the square root of the spot area is a linear function of the logarithm of the weight of material contained in it. Quantitative analysis is possible by relating the area of the sample spot to the plot of a series of standards.

The spots can be scraped from the plate and eluted with a solvent prior to quantitation with a suitable detector. In this case, colorimetric reactions may be carried out on eluted solutions, or UV or fluorometric analysis may be carried out.

Instruments that can provide direct quantitation are listed here: Radioscanners, or densitometers that can measure transmittance, reflectance, or fluorescence. Because of the unique nature of radioactive emissions, radioscanners work well with radioactive samples.

It should be noted that transmittance measurement is useful for visible light. Special steps have to be used for UV measurements since glass plates absorb UV light. If direct transmittance is preferred when UV detection is being utilized, then quartz plates have to be used or other suitable steps must be taken. The reflectance technique works fairly well with suitable instrumentation. Fluorescence measurements often prove more useful.

VII. APPLICATIONS

Described here are some select applications for amino acids, pharmaceuticals, vitamins, and dyes. Numerous examples of separations by TLC are available in scientific literature. The examples described below show a range of separations that are possible with this technique.

A. Amino Acids

A number of amino acids have been resolved by TLC. Table 3 lists the TLC systems that have been used with various amino acids.[4] The plates are developed to a distance of 10 cm.

The mobile phases A–G given in Table 3 show the extent of separations that are possible by utilizing a relatively "neutral" mobile phase such as ethanol/water or propanol/water, as opposed to using a basic mobile phase (see mobile phase E) or when phenol is used instead of one of the aliphatic alcohols. The reader should remember to consider the difference in the ratio of solvents prior to making any absolute judgments. It is interesting to note that in neutral solvents, arginine and lysine migrate much more slowly than the other amino acids. This difference has been attributed to cation exchange. Better separations are noted with System C and System G.

B. Pharmaceuticals

Sulfonamides are well-known antibacterial agents. They can be resolved on Whatman LHPKF layer with a mobile phase of chloroform/methanol (85:15). The development is done to a distance of 5 cm, and the samples are detected by UV quenching. The detection limit is below 50 ng, and quantitation is possible with a Kontes Model 800 densitometer with a 254-nm filter. The R_f values of three sulfonamides separated this way are:[1]

- Sulfathiazole   0.66
- Sulfapyridine   0.56
- Sulfamethazine   0.45

Another example involves determination of cholesterol in blood plasma.[1] The suitably diluted plasma is applied on the preadsorbent. The separation is achieved on Whatman LHPK with a chloroform/ethyl acetate (94:6) system, and the plate is developed to a distance of 6.5 cm. Detection is performed by spraying the plate with 3% cupric acetate in 8% phosphoric acid and heating the plate at 180 °C for 10 minutes. An R_f value of 0.44 is obtained for cholesterol. Quantitation is accomplished with a Kontes Model 800 using white phosphor. Detection limits of 10 ng have been reported.

A capsule dosage form containing phenylbutazone (see Fig. 5) and alkalizers showed a large increase in prednisone assay values with newly developed blue tetrazolium (BT) spray reagent under accelerated conditions.[5] The spray reagent is prepared by mixing 0.35% blue tetrazolium solution in 95% ethanol with 5 N sodium hydroxide in a ratio of 2:1. An investigation using TLC revealed the presence of four impurities (II, III, IV, and VI in Fig. 5); none of these produce color with BT reagent. Ahuja and Spitzer[6] developed the BT spray reagent (described above) and a TLC system (cyclohexane, chloroform, and ammonium hydroxide are mixed in the ratio of 40:50:10; the mixture is shaken and the organic phase of the resultant two-phase system is used) to track the degradation product(s) responsible for the problem. Two new transformation products have been characterized (V and VII). Compound VII was found to be responsible for interference in the prednisone assay.

TABLE 3 $R_f \times 100$ Values of Amino Acids in Various Solvents on Silica Gel G

Solvent	A	B	C	D	E	F	G
α-Alanine	47	37	27	39	40	49	25
β-Alanine	33	26	27	30	29	49	20
α-Aminobutyric acid						54	25
β-Aminobutyric acid						48	32
γ-Aminobutyric acid						38	30
α-Aminoisobutyric acid						57	26
β-Aminoisobutyric acid						4:6	29
α-Aminocaprylic acid	66	65	60	58	60		
Arginine	4	2	8	10	6	6	14
Asparagine						46	19
Aspartic acid	55	33	21	9	7	56	5
Citrulline						46	29
Cysteic acid	69	50	14	17	21	61	5
Cystine	39	32	16	27	22	8	9
Dihydroxyphenylalanine			45				
Glutamic acid	63	35	27	14	15	55	7
Glutamine						55	28
Glycine	43	32	22	29	34	50	18
Histidine	33	20	6	38	42	33	24
Hydroxyproline	44	34	20	28	31	63	33
Isoleucine	60	53	46	52	58	60	36
Leucine	61	55	47	53	58	63	37
Lysine	3	2	5	18	11	5	8
Methionine	59	51	40	51	60	62	36
Norleucine	61	57	49	53	59	66	54
Norvaline	56	50	38	49	57	67	56
Ornithine						6	5
Phenylalanine	63	59	49	54	60	63	41
Proline	35	26	19	37	30	48	45
Sarcosine	31	22	17	34	31		
Serine	48	35	22	27	31	52	19
Taurine						59	22
Threonine	50	37	2.5	37	40	66	18
Tryptophan	65	62	56	55	58	69	45
Tyrosine	65	62	56	55	58	65	36
Valine	55	45	35	48	56	56	29

Note: Adapted from Reference 4.
A = 96% Ethanol/water (7:3)
B = n-Propanol/water (7:3)
C = n-Butanol/acetic acid/water (4:1:1)
D = n-Propanol/34% ammonium hydroxide (7:3)
E = 96% Ethanol/34% ammonium hydroxide (7:3)
F = n-Propanol/water (1:1)
G = Phenol/water (3:1)

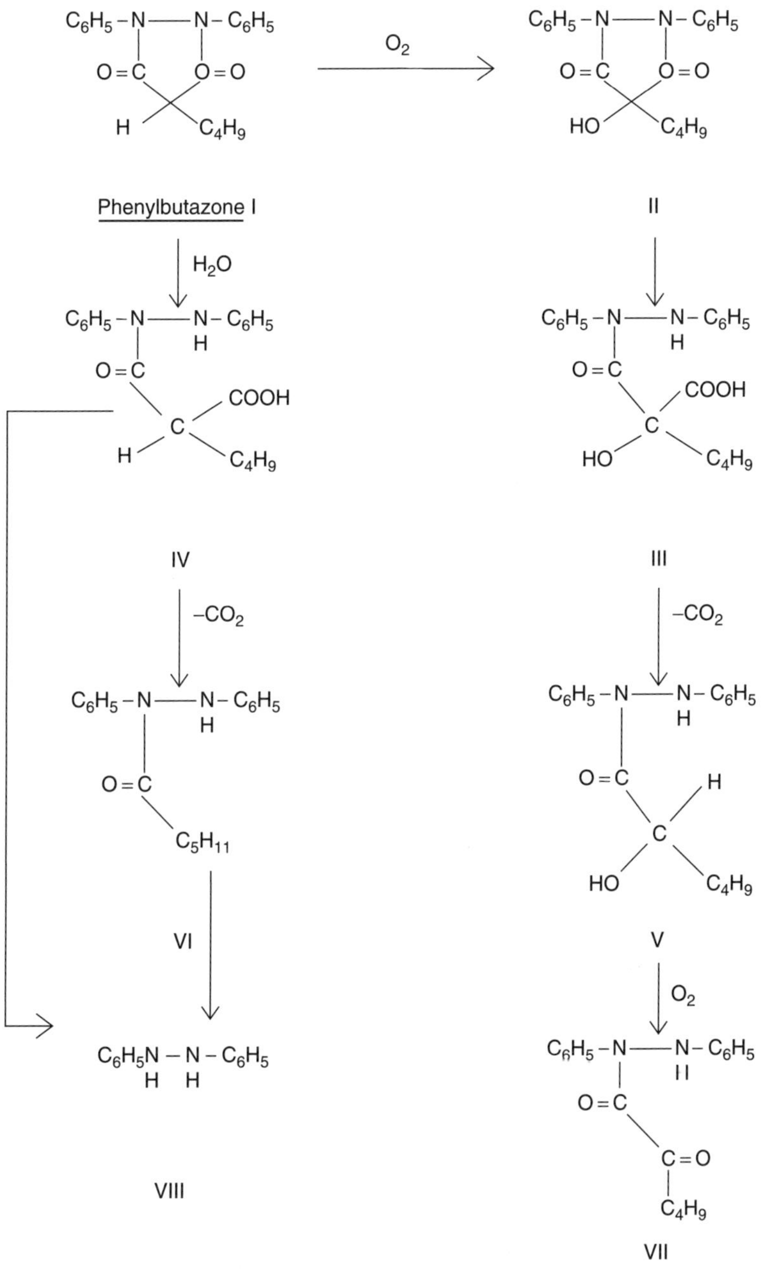

8 THIN-LAYER CHROMATOGRAPHY

FIGURE 5 Phenylbutazone degradation.

Two capsule formulations containing homatropine methylbromide (HMB), antacids, and other ingredients have been investigated for HMB stability under accelerated conditions.[7] TLC was performed with an acetic acetate/ethyl acetate/water/hydrochloric acid system (35:55:10:20) on a silica gel plate. Spraying the developed TLC plate with Dragendorff reagent revealed the presence of a purple spot. The purple spot is attributable to degradation of HMB to

TABLE 4 $R_f \times 100$ Values of Some Water-Soluble Vitamins

Vitamin	Silica gel G	Alumina
$B_1 \cdot HCl$	0	54
B_2	29	24
$B_6 \cdot HCl$	12	26
Biotin	55	54
Calcium pantothenate	40	0
Nicotinamide	44	62
Ascorbic acid	25	0
Folic acid	7	0
B_{12}	0	23
Carnitine HCl	3	20
Inositol	2	0

Note: Adapted from References 9 and 10.
Mobile phase: $HOAc/CH_3COCH_3/MeOH/C_6H_6$ (5:5:20:70)

methyltropinium bromide or tropine methylbromide (TMB). This suggests that the Dragendorff reaction can be useful for monitoring the stability of HMB.

Digitalis glycosides have been resolved on a reversed-phase plate (Whatman LK C-18). The mobile phase is methanol/water (70:30), and the plate is developed to a distance of 13 cm.[8] The plates are sprayed with chloramine T-trichloroacetic acid and viewed under UV at 366 nm. The R_f values of various glycosides are as follows:

- Digoxigenin monodigitoxoside   0.72
- Digoxigenin bisdigitoxoside   0.65
- Digoxin   0.60
- Gitoxin   0.45
- Digitoxin   0.35

C. Vitamins

The separation of a few water-soluble vitamins is shown in Table 4.[9,10] The data shows the difference in selectivity offered by the adsorbents; for example, vitamins B_1 and B_{12} are not resolved on silica gel but are well resolved on alumina. It should also be noticed that a number of vitamins are not resolved on alumina. This suggests the need to appropriately select a suitable stationary phase or to properly modify the mobile phase to achieve the desired separation.

D. Dyes

Table 5 shows the separation of a number of oil-soluble dyes on three different adsorbents.[11] Single solvents such as chloroform and xylene have

TABLE 5 $R_f \times 100$ of Some Oil-Soluble Dyes*

Dye	Silica gel G	Alumina	Kieselguhr
Martius yellow	28	0	0
Butter yellow	68	59	85
Sudan I	77	56	63
Sudan G	36	0	0
Ceres red G	30	19	16

*Adapted from Reference 11.
Mobile phases for silica gel G: petroleum ether/ether/acetic acid (70:30:1); aluminum oxide G: hexane/ethyl acetate (98:2); kieselguhr: cyclohexane

also been used for silica gel; however, mixed solvents frequently give better results.

REFERENCES

1. Touchstone, J. C. *Ultratrace Analysis of Pharmaceuticals and Other Compounds of Interest* (S. Ahuja, Ed.), Wiley, New York, NY, 1986.
2. Stahl, E. and Mangold, H. K. *Chromatography* (E. Heftman, Ed.), Van Nostrand, Reinhold, 1975; Stahl, E. *Thin Layer Chromatography*, Springer-Verlag, New York, NY, 1969.
3. Ahuja, S. *Trace and Ultratrace Analysis by HPLC*, Wiley, NY, 1992.
4. Brenner, M. and Niederweiser, A. *Experientia* 16:378 (1960).
5. Ahuja, S. and Spitzer, C. personal communication, October 27, 1966.
6. Ahuja, S. *Techniques and Applications of Thin Layer Chromatography* (J. Touchstone and J. Sherma, Eds.), Wiley, New York, NY, 1985, p. 109.
7. Spiegel, D., Ahuja, S., and Brofazi, F. R. *J. Pharm. Sci.* 61:1630 (1972).
8. Block, D. E. *J. Assoc. Off. Anal. Chem.* 631:707 (1980).
9. Gaenshirt, H. and Malzacher, A. *Naturwiss.* 47:279 (1960).
10. Katsui, G., Ishikawa, S., Shimizu, M., and Nishimoto, Y. *Chem. Abstr.* 60:9577 (1964).
11. Pereboom, J. W. C. *Chem. Weekbl.* 57:625 (1961).

QUESTIONS FOR REVIEW

1. Define TLC. What advantages does it offer over paper chromatography?
2. List various modes of TLC.
3. How does normal-phase TLC differ from reversed-phase TLC?
4. Provide some universal detection reagents for TLC.

9
GAS CHROMATOGRAPHY

    I. EQUIPMENT
    II. SEPARATION PROCESS
    III. COLUMNS
        A. Solid Support
        B. Stationary Phases
    IV. GAS CHROMATOGRAPHIC DETECTORS
        A. Detectability
        B. Design of Commonly Used Detectors
    V. RECORDING AND ANALYSIS
    VI. RESOLUTION
    VII. SELECTION OF A STATIONARY PHASE
        A. Sample Classification
        B. Suitable Stationary Phases
        REFERENCES
        QUESTIONS FOR REVIEW

In gas chromatography, one of the phases is gas and the other is generally a solid or a liquid retained on a solid support. Chemically bonded stationary phases are also commonly used (see discussion below). When a liquid phase is used as a stationary phase or when a solid coated onto a solid support becomes liquid at the operational temperatures, this technique is more appropriately called gas–liquid chromatography (GLC). And, by corollary, where only a solid stationary phase is used, the technique should be designated gas–solid chromatography (GSC). Unfortunately, both of these techniques are frequently simply called gas chromatography or GC (see discussion in the following paragraph).

As mentioned earlier, in GLC, the stationary phase is a liquid or a solid that becomes liquid at the operational temperature. The liquid phase is virtually immobilized by coating it onto a solid support that should be inert for all practical purposes. Instead of coating a liquid on a support, it is not uncommon to carry out select chemical reactions with the solid support, with the expectation that the resulting phase will behave somewhat like a liquid phase. However, it must be recognized that the resulting stationary phase is not liquid. This may partially explain why the stationary phase is generally not mentioned in the designation of this technique and it is simply called gas chromatography (GC). The term GC is also used, albeit incorrectly, when a solid adsorbent is used as a stationary phase or when a material is reacted with the solid support to produce a stationary phase that is less likely to bleed from the column. It would be more appropriate to describe this form of chromatography as gas–solid chromatography. However, instead of arguing with this traditional misuse of terms, we will go along with most practitioners and use the term GC interchangeably with GLC as long the difference between the two techniques is well understood.

Historically, the origin of gas chromatography can be dated back to 1905 when W. Ramsey separated a mixture of gases and vapors from solid adsorbents such as activated charcoal. This technique entailed adsorption and can be described as gas–solid adsorption, a forerunner of gas–solid chromatography that preceded the discovery of liquid–solid column chromatography by Tswett in 1906 (see discussion in Chapter 1). Tswett is generally considered the father of chromatography because his work with column chromatography led to significant additional advancements in this new technique. Furthermore, he designated this technique *chromatography* (which, literally translated from Greek, means color writing). It is well known that chromatography performed today is not restricted to colored compounds or to gravity-fed columns. Modern chromatography embraces powerful resolution techniques that include gas chromatography, high-pressure liquid chromatography (see Chapter 10), supercritical fluid chromatography, and capillary electrokinetic chromatography (see Chapter 11).

Gas chromatography, where gas is used as a mobile phase, was first introduced in 1952 by James and Martin.[2] The technique was based on a suggestion made 11 years earlier by Martin in a study by Martin and Synge[3] on partition chromatography, for which they were presented the Nobel Prize in chemistry in 1952. As previously mentioned, the stationary phase can be a solid or a liquid immobilized on solid; the phase remains in the column. Both packed and open tubular columns can be used (see Section III). The observed separations in gas chromatography are based on the relative vapor pressures of sample components and affinities for the stationary phase.

The sensitivity, speed, accuracy, and simplicity of the gas chromatographic methods for the separation, identification, and quantification of volatile compounds or those that can be made volatile under gas chromatographic conditions has resulted in phenomenal growth of this technique.[4,5]

I. EQUIPMENT

The basic components of a gas chromatograph are shown in Figure 1. It includes a cylinder of carrier gas equipped with a flow controller and pressure regulator. Helium is most frequently used as a carrier gas. Nitrogen or argon may also be used. The primary requirement is that the gas be inert and not interfere with the detection of the desired component(s). The carrier

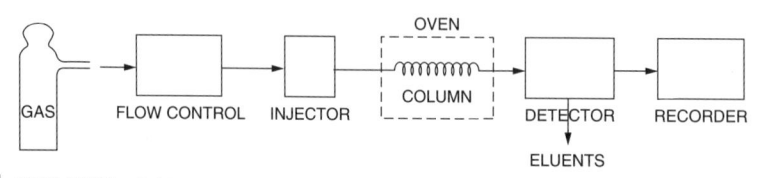

FIGURE 1 A block diagram of a gas chromatograph.

gases commonly used for various frequently used detectors are given below.

Detector	Carrier gas
Thermal conductivity	Helium
Flame ionization	Helium or nitrogen
Electron capture	Nitrogen or argon

For electron-capture detection, the nitrogen used should be extremely dry. Carrier gases are available at various purity levels that range from 99.995% to 99.9999%. The purity of the gas used relates to the nature of experimentation.

The carrier gas flows through an injection port onto a column, which is in turn connected to a detector. The gas exits into the atmosphere from the detector.

Gas, liquid, or solid samples can be introduced into the gas chromatograph. The sample is injected through the injection port into the dynamic system at the head of the column, depending on the type of column, as shown below. In packed columns, the sample inlet allows flash vaporization or on-column injections have to be made. Ideally, the sample should be injected instantaneously on-column, but in practice this is impossible; therefore, an effort is made to introduce the sample rapidly to produce a symmetrical band.

In capillary columns, the sample inlets allow the following modes of injection:

- Split
- Splitless
- On-column

To prevent overload of the capillary columns, it is often desirable to split the sample injected into a gas chromatograph in a predetermined ratio such that only a small portion of the sample is injected onto the column. Splitless injection entails injection of accurately measured small volumes of the sample. On-column injections are generally preferred in gas chromatography because they minimize band broadening.

The column contains a stationary phase that allows separations of sample components to occur, which in turn can be detected by the detector. A number of different detectors can be used. For example, if we want to detect all samples without undue concern for low-level detectability, a thermal conductivity detector can be used. The thermal conductivity detector is a simple, virtually universal detector that measures the changes in thermal conductivity. The injector, column, and detector are maintained at temperatures that are thermostatically controlled, to allow an efficient separation to occur. Further details on the chromatographic system follow.

II. SEPARATION PROCESS

The separation process is described here (in other forms of gas chromatography, selectivity offered by the stationary phase can be substituted for the liquid phase in GLC).

In GLC, the sample components to be separated are carried by the inert carrier gas through the column. The sample components distribute between the carrier gas and a relatively nonvolatile liquid (stationary phase) supported on an inert, size-graded solid (solid support). The stationary phase selectively retards the progress of sample components according to their distribution coefficient such that they form separate bands in the carrier gas. These component bands leave the column in the stream of carrier gas and are recorded as a function of time by the detector.

The main advantages of GLC are as follows:

1. It is reliable, relatively simple, and inexpensive compared to other modern chromatographic techniques.
2. The analysis time is short.
3. The sample components are completely separated into bands (which can be recorded as peaks) in the carrier gas. This makes quantification or collection of fractions easy when a nondestructive detector is used.
4. The column is continuously regenerated by the carrier gas.
5. It is an efficient technique that provides high resolution of components.
6. Low-level detectability is possible.
7. Accurate analysis with relative standard deviation (RSD) of 1% to 5% is possible.
8. Small samples, typically in microliter volume, can be analyzed.

The primary disadvantages are as follows:

1. The sample components have to be volatile at the operational temperature, or volatilization must be achieved by suitable derivatization techniques.
2. The strongly retained components travel very slowly, or in some cases do not move at all. This problem generally can be overcome by temperature programming of the column to decrease elution time. (Temperature programming constitutes increasing the column temperature at some finite rate during the analysis to provide a faster and more versatile analysis.)
3. It is not ideally suited as a large-scale preparative technique.
4. Thermally labile samples have to be derivatized prior to chromatography.

III. COLUMNS

After the appropriate hardware listed under Equipment (Section I) has been selected, it is necessary to select a suitable column with an appropriate stationary phase to complete a good chromatographic system. The column can be considered the heart of a chromatographic system. It is generally composed of tubing of stainless steel or glass that contains the stationary phase. Aluminum and copper tubing have also been used; however, copper tubing should be avoided because of its reactivity.

The column can be straight or coiled to fit the gas chromatograph. If columns are coiled, the spiral diameter should be at least 10 times the column-tubing diameter to minimize diffusion and racetrack effects. The column-tubing diameter varies from 0.01 inch up to more than 2-inch internal diameter (i.d.). Smaller column diameters generally lead to better efficiency. Standard analytical columns are 1/4-inch outer diameter (o.d.). Higher column capacity requires larger column diameter.

The column is packed with solid support that has been coated with the stationary phase, or the walls of the capillary columns are coated with the stationary phase. Support-coated lower-diameter columns can also be used. Columns can be divided into two categories:

- Packed columns
- Open tubular columns

Packed columns are commonly used in routine GC and GLC, their outer diameter is 1/4 inch or 1/8 inch, and they may be three to twelve feet in length. Selection of solid support and stationary phase in regard to columns is discussed below.

Open tubular columns are capillary columns that have an inside diameter of 0.25 mm and are not packed. Long lengths up to 100 m can be used. These columns offer very efficient separations for complex sample mixtures. They can be further divided as follows:

- Wall-coated open tubular (WCOT) columns
- Support-coated open tubular (SCOT) columns
- Porous-layer open tubular (PLOT) columns

A. Solid Support

The object of using solid support is to provide a large, uniform, inert surface area for spreading the liquid stationary phase. Some of the desirable properties of solid support are listed below.[4] It should

- be relatively inert, i.e., it should show minimal adsorption
- have large surface area
- have uniform size and shape
- be resistant to crushing

There are two basic solid support products with the trademark Chromosorb that are commonly used in GC. Chromosorb P (P stands for the pink color) is prepared from C-22 firebrick, while Chromosorb W (W stands for white) is prepared from celite filter aids. Chromosorb P is used when high column efficiency is required; however, it shows strong adsorption for polar compounds. By contrast, Chromosorb W provides a relatively inert surface and is suitable for polar compounds; the column efficiency is not as high as that obtained with Chromosorb P. A fluorocarbon polymer called Chromosorb T with much larger surface area (7.5 m^2/g) is also available as solid support. Its packed density is 0.42 g/cc and allows a maximum loading of 10% for the liquid phase.

Some of the diatomaceous solid supports that are manufactured and trademarked by Celite Corporation are given in Table 1. Diatomaceous earth surfaces are generally too active to allow good separation of polar compounds.

TABLE I Diatomaceous Solid Supports

Name	Surface area (m²/g)	Packed density (g/cc)	Pore size (μm)	% Loading (maximum)
Chromosorb P	4.0	0.47	0.4–2	30
Chromosorb W	1.0	0.24	8–9	15
Chromosorb G	0.5	0.58	Not available	5
Chromosorb 750	0.7	0.40	Not available	7

Source: Celite Corp.

They contain free hydroxyl groups that can form undesirable hydrogen bonds to solute molecules and cause peaks to tail. Even the most inert material, Chromosorb W, needs to be acid-washed (Chromosorb AW) and then silanized to make it less active and thus more inert.

Some of the commonly used silanizing reagents are dimethyldichlorosilane (DMDCS) and hexamethyldisilazane (HMDS). The deactivated supports are sold under various names:

- Supelcoport
- Chromosorb W-HP
- Gas Chrom Q II
- Anachrom Q

The primary disadvantage of deactivating these supports is that they become too hydrophobic. As a result, it is difficult to coat them with a polar liquid phase.

B. Stationary Phases

Both solid and liquid stationary phases that are commonly used in gas chromatography are discussed below.

1. Solid Stationary Phases

Adsorbents like silica gel and alumina can be used in GSC. A number of specific stationary phases are also available that have been developed specially for GSC (see Table 2).

A separation of gases such as hydrogen, carbon monoxide, methane, ethane, and carbon dioxide is possible on approximately 20 feet of silica gel column at 60 °C. However, this column cannot separate oxygen and nitrogen. The separation of oxygen and nitrogen, as well as helium, argon, and methane, has been achieved with molecular sieves on a 25-m PLOT column.[4] An excellent separation of various volatile solvents is possible on Carbopack column. Porapak commercial columns made of porous polymer have found good use in determination of water in polar solvents.

2. Liquid Phases

A solid support coated with a liquid phase is a simple example of a stationary phase. A solution of the liquid phase is made in a volatile solvent. The solid

TABLE 2 Common GSC Adsorbents

Common name	Commercial names
Alumina, activated	Alumina F-1, Unibeads A
Silica gel	Chromasil, Davidson Grade 12, Porasil
Molecular sieves, zeolite	MS 5A, MS 13X
Molecular sieves, carbon	Carbopack, Carbotrap, Carbograph, Graphpac
Polymers, porous	Porapak, Hayesep, Chromosorb Century Series
Polymer, Tenax	Tenax TA, Tenax GR

Note: Adapted from Reference 4.

support is added to it, mixed well, and then the solvent is removed by evaporation. These materials appear dry even at fairly high loading and can be easily packed in a column. Low liquid-phase loadings yield high efficiency and are more useful for high-boiling compounds, and high loadings are useful for large sample size or volatile solutes such as low boilers and gases.

The importance of the stationary phase cannot be overemphasized in gas chromatography. A good stationary phase in GLC should have the following characteristics:

- There should be negligible vapor pressure at operating temperature.
- The phase should have good stability at operating temperature.
- Samples should have reasonable solubility in the phase.
- Sample components should exhibit different distribution coefficients.

An impressive advantage of gas chromatography is that substances with the same vapor pressure can be separated by appropriate selection of the stationary phase. The versatility and selectivity of gas chromatography lies in the wide choice of stationary phases. As a matter of fact, the choice of stationary phases may be too wide. Over 200 of them are available in various vendor catalogues. Several chromatographers have attempted to shorten this list.[4] The following stationary phases have been found useful over the years:

- OV-1
- OV-17
- Carbowax 20M
- DEGS
- OV-210
- OV-275

These stationary phases cover a wide range, from nonpolar methyl silicones to a highly polar silicone like OV 275, and a polyhydroxy compound like Carbowax 20M. Carbowaxes with molecular weights as low as 1500 may be found useful for some samples. For more information on selectivity of these stationary phases, see Section VII. B.

It is not easy to quantify the polarity of a stationary phase in gas chromatography. The polarity is determined by intermolecular forces, which are quite complex and often difficult to predict in a chromatographic system. Intermolecular interactions can play a significant role in gas chromatography (see earlier discussion on this subject in Chapter 4). The dipole moment of a pure liquid indicates its polarity. For example, a large dipole moment and high boiling point would suggest that the liquid is highly polar and strong intermolecular forces are at play.

3. Intermolecular Interactions

In GC, we are interested in intermolecular forces between the solute molecules in vapor state and stationary phase in liquid state. The following important intermolecular interactions (see Chapter 4) should be given special consideration prior to the selection of a stationary phase.

- Orientation interactions
- Debye interactions
- Dispersion interactions
- Specific interactions

The first three interactions have also been called van der Waals forces.

a. Orientation interactions

The interactions between two permanent dipoles are called orientation interactions; they are also called dipole–dipole interactions or Keesom forces.

Hydrogen-bond formation is an important orientation interaction in gas chromatography. It may be recalled from the discussion in Chapter 4 that an atom of hydrogen is attracted by two atoms instead of only one, under certain conditions, such that it may be considered to be acting as a bond. The hydrogen atom is rather unique in that when its electron becomes detached, only the nucleus of a very small diameter is left. Hydrogen bonding becomes readily evident if one of the molecules has a hydrogen atom bonded to an electronegative atom like nitrogen or sulfur. For example, alcohols and amines can donate or receive a hydrogen atom to form a hydrogen bond.

It is frequently necessary to consider hydrogen bonding for compounds containing –COOH, –OH, and –NH$_2$ groups. In these cases, a proton-donor group (e.g., –OH, >NH, –SH, HCCl$_3$) interacts with a proton acceptor, an atom or group with unpaired electrons (e.g., –O–, =N–, –F, –S–, –Cl, >C=C<, phenyl). The energy of the resulting interaction is usually large when the two atoms attached to the bonding proton are nitrogen or oxygen. The basis of hydrogen bonding is largely electrostatic, the normally small O–H or N–H dipole being effectively magnified by the small radius of the hydrogen atom. Hydrogen bonding can help provide some very useful separations in gas chromatography.

Hydrogen bonding can also cause some undesirable interactions. For example, solutes that are capable of hydrogen bonding can become attached to the solid support or walls of the injection port or the column tubing. This results in slow desorption of solute from the column, which can give rise to asymmetrical peaks. Derivatization of surface hydroxyl groups of affected areas can minimize these effects.

b. Debye interactions

The interactions between a permanent dipole in one molecule and the induced dipole in a neighboring molecule are described as Debye interactions. These interactions can play some role in terms of selectivity of separation in gas chromatography of the components of interest. These forces are usually very small.

c. Dispersion interactions

The forces arising out of synchronized variations in the instantaneous dipole of the two interaction species are called dispersion interactions. Frequently, they are called London forces. These forces are present in all cases and are the only source of attractive energy between two nonpolar substances.

Dispersion interactions are weaker than the induction and orientation forces discussed earlier and are significant mainly for nonpolar solutes. Induction and orientation forces can provide selectivity in chromatographic systems and may be responsible for the observed polarity.

d. Specific interactions

Specific interactions (see earlier discussion in Chapter 4) relate to forces resulting from chemical bonding or complex formation between the solute and stationary phase. The forces resulting from these interactions determine solubility of the solute and thereby affect the desired separation.

The combined effect of these interactions can be expressed by the partition coefficient k as follows:

$$k = \frac{\text{amount of solute per unit volume of stationary phase}}{\text{amount of solute per unit volume of gas phase}}$$

The value of k is high when a larger amount of an analyte (solute of interest) is retained in the stationary phase. This means that the analyte moves slowly down the column because only a small fraction is in the carrier gas at any given time. The movement of the solute is negligible in the liquid phase, and only the fractions that partition back into the gas phase move down the column.

In short, separation between two compounds in gas chromatography is possible if their partition coefficients are different. The greater difference in the k values makes separation easier. In the next section, we will discuss how this can translate into use of a smaller number of theoretical plates or a shorter column. Further discussion on selection of stationary phases is given in Section VII.

e. Column efficiency

The resolution of complex mixtures is frequently controlled by column efficiency given by N, the number of theoretical plates in a column or by H, the height equivalent of a theoretical plate.

4. van Deemter Equation

Earlier in this book, a general form of the van Deemter equation was described. This equation is given here in a more detailed form. It was originally developed for GC; however, it has been found useful for all forms of chromatography including those that evolved later than GC, such as HPLC and SFC. The equation for a packed column is as follows:

$$H = 2\lambda d_p + \frac{2\lambda D_M}{u} + \frac{8k' d_f^2 u}{\pi^2(1+k')^2 D_S} + \frac{\omega d_p^2 u}{D_M}$$

$$A \text{ term} \quad B \text{ term} \quad C_{liquid} \text{ term} \quad C_{gas} \text{ term}$$

where λ, γ, and ω are functions of column packing, C_{liquid} and C_{gas} terms contribute to mass transfer effects, d_f^2 = diffusion thickness of liquid film, D_S = diffusion coefficient in the stationary phase, D_M = diffusion coefficient in the mobile phase, d_p = particle diameter, k' = capacity factor.

Several modifications of this equation have been proposed to account for various effects encountered in chromatography; however, for most practical purposes the equation as shown above is adequate.

IV. GAS CHROMATOGRAPHIC DETECTORS

A large number of detectors have been used in gas chromatography; many of these detectors, except the thermal conductivity detector, were developed specifically for GLC. Some of these detectors are listed in Table 3. In general, they can be classified on the basis of whether they can be used for all compounds or whether they are selective for a given class of compounds.

The characteristics of three detectors most commonly used are as follows:

- Thermal conductivity detector (TCD)—it can be considered a universal detector that responds to concentration and bulk property; it is nondestructive.

TABLE 3 Applications of Commercially Available Detectors for GC

Detector	Applications
Thermal conductivity (TCD)	Universal
Flame ionization (FID)	Most carbon compounds
Electron capture (ECD)	Halogenated compounds
Photoionization (PID)	Aromatic compounds
Nitrogen/phosphorus (NPD)	N-, P-, and halogen-containing compounds
Flame photometric (FPD)	S- and P-containing compounds
Atomic emission (AED)	Metals; halogens, C- and O-containing compounds
Electroconductivity (ECD)	S-, N-, and halogen-containing compounds
Chemiluminescent	S-containing compounds
Radioactivity	^{3}H- and ^{14}C-containing compounds
Mass spectrometer (MSD)	Variety of compounds

Note: Adapted from Reference 4.

- Flame ionization detector (FID)—it is selective for carbon compounds, responds to mass flow-rate and specific property, and is destructive.
- Electron capture detector (ECD)—it is selective for halogen compounds, responds to bulk property, and is nondestructive.

The speed of response of a detector is important. This is generally referred to as the time constant. Specifically, it is the time, in milliseconds, that a detector takes to respond to 63.2% of sudden change in signal. The response time corresponds to 98% of full response and is equal to 4 times the time constant. It is recommended that the time constant be less than 10% of the peak width at half height.

A. Detectability

The minimum amount detected (MAD) or minimum limit of detection (MLD) or, more simply, limit of detection (LOD) is generally considered to be a detector response that is at least twice the noise level or baseline response. In general, it is preferable to determine this value practically.

Thermal conductivity detectors can detect levels of an analyte down to 100 ppm. It is virtually a universal detector that operates on measuring the difference in conductivity in sample cell versus a blank, that is, the carrier gas.

The flame ionization detector can evaluate low parts per million of most organic compounds. The sample on exiting the column is exposed to flame, causing ionization of resulting compound(s) to occur, which are then detected. Some compounds, e.g. ringed nitrogen compounds, show poor response with this detector.

Specific detectors, such as electron capture detectors, can determine parts per billion, or picograms, of selected halogenated compounds. A number of other specific detectors such as photoionization detector, flame photometric detector, nitrogen phosphorous detectors are also available. Mass spectrometer is a universal detector that is commonly used in gas chromatography.

B. Design of Commonly Used Detectors

The structure of the three commonly used detectors is given here in the schematic form. A brief description of these detectors has also been included.

I. Thermal Conductivity Detector

As mentioned before, a thermal conductivity detector (TCD) is a differential detector that measures the difference in thermal conductivity of the sample in carrier gas as compared to the thermal conductivity of carrier gas alone (blank). In a conventional detector, at least two cells are used. A typical TCD cell is shown in Figure 2. Small cell volume is very desirable. TCD cells with volume around 140 µl are useful for packed columns or wide-bore capillaries.

Higher sensitivity is obtained by applying a larger heating current to the filaments; this results in greater temperature differential and sensitivity. However, high temperatures shorten filament life because of oxidation due to small amounts of impurities, resulting in ultimate burnout of the tungsten wire.

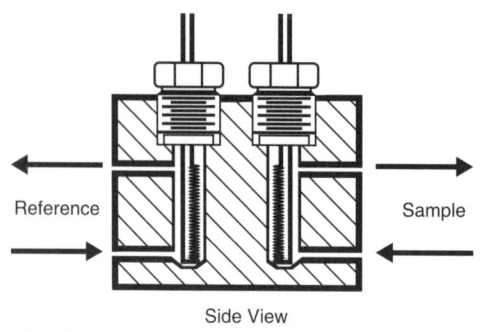

FIGURE 2 A typical TCD cell. (Courtesy of Gow–Mac Instruments)

The carrier gas used with the thermal conductivity detector must have thermal conductivity that is very different from the samples to be analyzed. Helium and hydrogen are the recommended carrier gases, and helium is the gas more commonly used.

2. Flame Ionization Detector

The flame ionization detector (FID) is the most commonly used detector in GC. The column effluent is burned in a small oxygen–hydrogen flame to produce some ions. When collected, the ions produce a small current that provides the desired signal. When no sample is being burned, there should be minimal ionization and thus the blank signal is very small.

A typical FID is shown in Figure 3. The column eluent is mixed with hydrogen and leads to a small burner that is surrounded by a high flow of air

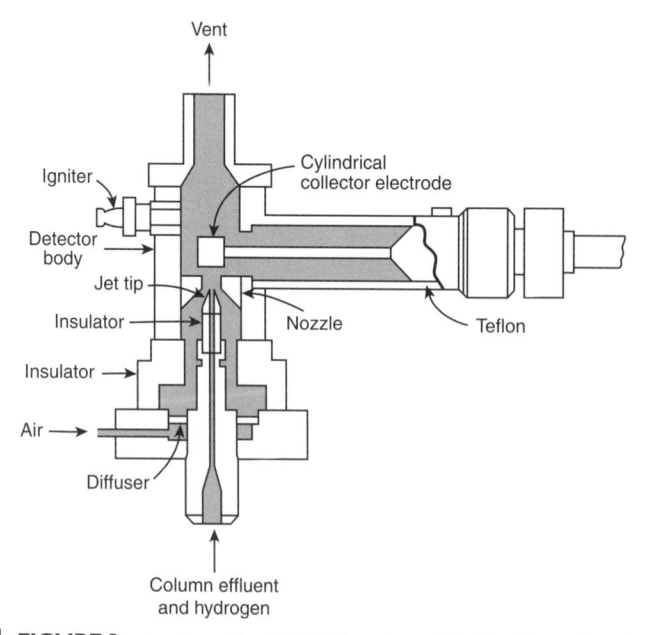

FIGURE 3 A schematic of FID. (Courtesy of Perkin Elmer Corp.)

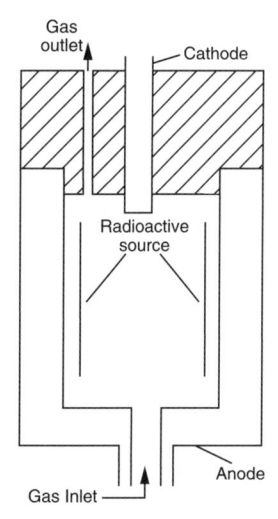

FIGURE 4 A schematic of ECD.

to support the combustion. The collector electrode is biased about $+300$ V relative to flame tip, and high impedance circuit amplifies the collected current. Since water is invariably produced in the combustion process, the detector must be heated at a sufficiently high temperature to prevent condensation of water. The response of FID roughly parallels the carbon content of the sample. The limitation of this detector is that compounds that do not contain organic carbon do not burn and are thus not detected.

3. Electron Capture Detector

It may be recalled that the electron capture detector (ECD) is a selective detector that provides very high response for those compounds that capture electrons. These compounds include halogenated materials such as pesticides. As a result, this detector is most commonly used for residue analysis of pesticides.

A typical ECD is shown in Figure 4. ^{63}Ni is generally used as the beta emitter, although tritium has also been used. Nickel is preferable because it can be used at high temperatures up to 400 °C and it is fairly safe to use.

ECD is an ionization type of detector but unlike most detectors of this class, samples are detected by causing a decrease in the level of ionization. It has been shown that improved performance is obtained if the applied voltage is pulsed rather than continuously applied. A square wave pulse of around -50 V is applied at a frequency to maintain constant current whether or not a solute is present in the cell; consequently, the pulse frequency is higher when no solute is present. The pulsed ECD improves the detection limits and provides a wider linear range to assist the quantification.

V. RECORDING AND ANALYSIS

Gas chromatography can be used for both qualitative and quantitative analyses. For qualitative analysis, retention time serves as a good index for identifying a component. Retention time is defined as time taken by a given component to elute from the column and is measured as time elapsed from the point of injection to the peak maximum. This property is characteristic of a sample and a liquid phase at a given temperature. It can be reproduced to an accuracy of better than 1% with appropriate control of temperature and flow rate.

The signal from the eluted components is conventionally recorded on a strip chart recorder, where time is the abscissa and the signal, in millivolts, is the ordinate. At present, the signal is more likely to be collected digitally and then plotted electronically with a computer.

The area of the peak for a given sample is proportional to its concentration. This allows exact determination of the content of a given component by calculation against a known standard at comparable concentration. The accuracy of quantification relates to sample concentration and preparation, injection, detection, and integration methods. An accuracy of better than 1% is possible with electronic digital integrators or computers.

VI. RESOLUTION

Separation between two peaks is determined by the resolution R_S of a given system. The resolution equation is as follows:

$$R_S = \frac{1}{4}\left(\alpha - \frac{1}{\alpha}\right)N^{1/2}\left(\frac{k'}{1+k'}\right)$$

where R_S = resolution, N = total number of theoretical plates in a column, α = separation factor, and k' = capacity factor.

It may be recalled that

$$\alpha = \frac{k'_2}{k'_1}$$

and

$$k' = \frac{\text{retention time of a peak} - \text{retention time of unretained peak}}{\text{retention time of unretained peak}}$$

In its simple form, the resolution equation can be described as follows:

$$R = \frac{t_2 - t_1}{(1/2)(W_1 + W_2)}$$

where t_2 and t_1 = retention times of peak 2 and peak 1, respectively and W_2 and W_1 = peak widths of peak 2 and peak 1, respectively.

The number of plates required for a given separation may be calculated from the resolution equation as follows:

$$N_{\text{req}} = 16R_S^2\left(\frac{\alpha}{\alpha-1}\right)^2\left(k'_2 + \frac{1}{k'_2}\right)^2$$

where N_{req} = number of plates required for a given separation and k'_2 = the peak capacity of the second peak.

9 GAS CHROMATOGRAPHY

VII. SELECTION OF A STATIONARY PHASE

The selection of a stationary phase is generally based on overall knowledge of the field, literature survey, and actual experimental work. A reliable stationary phase should meet the following criteria:[4]

- Good solubility for sample components because components with poor solubility elute rapidly.
- Good differentiation for sample components—this relates to their partition coefficient.
- Good thermal stability—instability can result from catalytic influence of solid support at high temperatures.
- Poor volatility; that is, the stationary phase should be nonvolatile and have a vapor pressure of 0.01 to 0.1 mm at operating temperatures.
- Poor reactivity with components of the sample at operating temperatures.

It is important to have at least some idea about the chemical nature of the sample component(s). If this information is not available, an intelligent guess is made as to the possible nature of the sample. Alternatively, it would be necessary to analyze a sample for all classes of compounds to assure that none of the components is missed. The basic plan is to try to classify the sample into one of five groups, on the basis of the polarity of its components.

A. Sample Classification

Samples can be divided into various groups, based on their polarity.[4]

1. Group 1 (polar)

This group contains compounds with both a donor atom, such as N, O, or F, and an active hydrogen atom. They include:

- Alcohols
- Fatty acids
- Phenols
- Amines (primary and secondary)
- Nitro compounds with α-H atoms
- Nitriles with α-H atoms

2. Group 2 (highly polar)

The compounds capable of forming a network of hydrogen bonds are included in this group. They include:

- Water
- Glycerol
- Amino alcohols
- Hydroxy acids
- Dibasic acids
- Polyphenols

3. Group 3 (intermediate polarity)

The compounds containing a donor atom but no active H atoms are included in this group.

- Aldehydes
- Ethers
- Esters
- Ketones
- Tertiary amines
- Nitro compounds with no α-H atoms
- Nitriles with no α-H atoms

4. Group 4 (low polarity)

The compounds with active H atoms but no donor atoms belong to this group.

- Halogenated hydrocarbons
- Aromatic hydrocarbons
- Olefins

5. Group 5 (nonpolar)

The compounds showing no hydrogen-bonding capacity are placed in this group.

- Saturated hydrocarbons
- Mercaptans
- Sulfides
- Halocarbons such as CCl_4

Suitable stationary phases for these samples are given in the next section.

B. Suitable Stationary Phases

Stationary phases can be similarly classified to assist selection for use with a given group of compounds.

1. Stationary Phases Suitable for Polar Compounds (Group 1)

- β,β-Oxydipropionitrile
- XE-60
- Tetracyanoethyl pentaerythritol
- Ethofat

2. Stationary Phases Suitable for Highly Polar Compounds (Group 2)

- Carbowaxes (various molecular weights)
- Carbowax 20M-TPA
- FFAP
- Diglycerol
- Castorwax

3. Stationary Phases Suitable for Compounds with Intermediate Polarity (Group 3)

- Dibutyl tetrachlorophthalate
- Tricresyl phosphate
- OV-17

4. Stationary Phases Suitable for Compounds with Low Polarity and for Nonpolar Compounds (Groups 4 and 5)

- SE-30
- Squalane
- Apiezons
- Hexadecane
- OV-1

The structures of some of the commonly used liquid phases are given below:[4]

$$HO-(CH_2-CH_2-O)_n-H$$

Carbowax

$$CH_3(CH_2)_5-\underset{OH}{CH}-CH_2-CH=CH-(CH_2)_{17}COOH$$

Castorwax

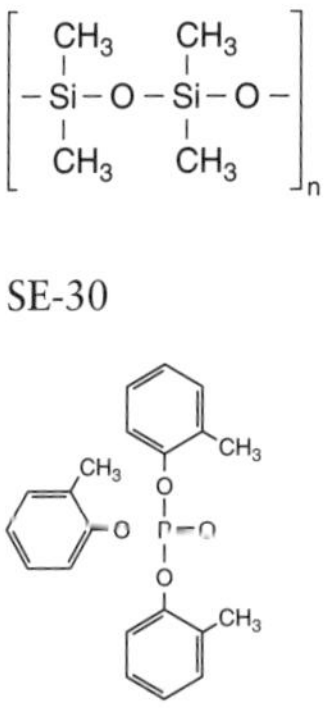

SE-30

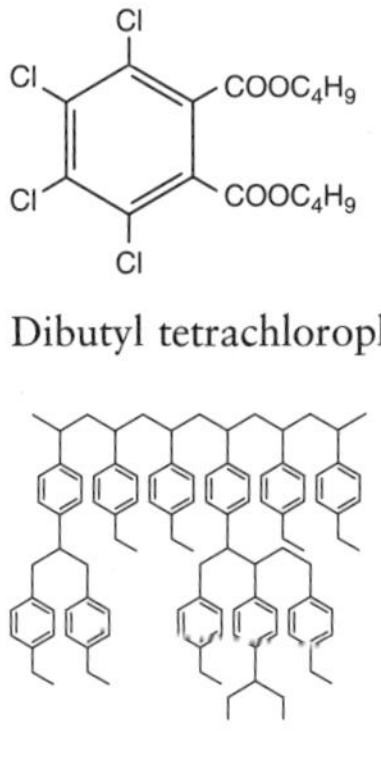

Dibutyl tetrachlorophthalate

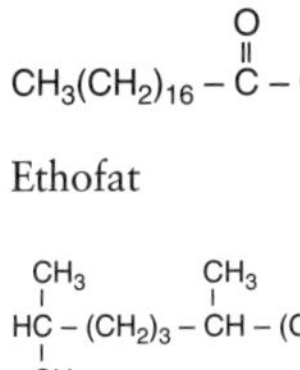

Tricresyl phosphate

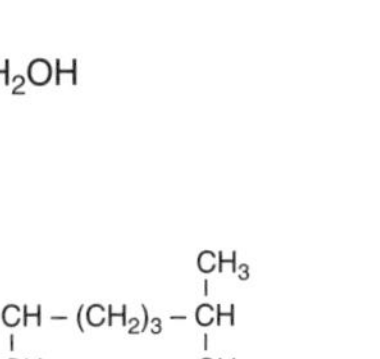

Porapak

$$CH_3(CH_2)_{16}-\overset{O}{\underset{\parallel}{C}}-O-(CH_2-CH_2-O)_n-CH_2CH_2OH$$

Ethofat

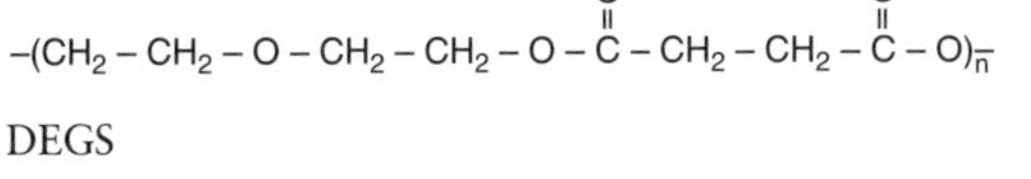

Squalane

$$-(CH_2-CH_2-O-CH_2-CH_2-O-\overset{O}{\underset{\parallel}{C}}-CH_2-CH_2-\overset{O}{\underset{\parallel}{C}}-O)_{\overline{n}}$$

DEGS

TABLE 4 Recommended Maximum Temperature for Operation of Liquid Phases

Liquid phase	Maximum temperature (°C)
Carbowaxes 400	125
Carbowax 1500	200
Carbowax 4000	200
Carbowax 20M-TPA	250
FFAP	275
Diglycerol	120
Castorwax	200
β,β-oxydipropionitrile	100
Silicone XE-60	275
Tetracyanoethylated pentaerythritol	180
Ethofat	140
Dibutyl tetrachlorophthalate	150
Tricresyl phosphate	125
OV-17	300
Silicone SE-30	300
Squalane	100
Apiezon J	300
n-Hexadecane	50
OV-1	350

The liquid phases should be operated under the recommended maximum temperatures (see Table 4).

5. Rohrschneider and McReynolds Constants

The Kovats index scale was advanced to define polarity of compounds. It uses a homologous series of n-paraffins as standards against which adjusted retention volumes are gauged for solutes of interest.

This approach was later replaced by a more elaborate system proposed by Rohrschneider. He developed a polarity scale for liquid phases, where a value of zero was assigned to the hydrocarbon squalane ($C_{30}H_{62}$) and a value of 100 was given to β,β-oxydipropionitrile.[6] Listed below are the five compounds proposed by Rohrschneider/McReynolds for characterizing any given liquid phase. Each is intended to measure a different type of solute–solvent interaction.

(a) Benzene
(b) Ethanol
(c) 2-Pentanone
(d) Nitromethane
(e) Pyridine

TABLE 5 Rohrschneider/McReynold's Constants for Characterizing Liquid Phases

Phase	Benzene	Ethanol	2-Pentanone	Nitromethane	Pyridine
Squalane	0	0	0	0	0
Apiezon L	35	28	19	37	47
SE-30	15	53	44	64	31
Di-*n*-octyl phthalate	96	188	150	236	172
Tricresyl phosphate	176	321	250	374	299
QF-1	144	233	355	463	305
OV-1	16	55[a]	44	65[b]	42
OV-17®	119	158[a]	162	243[b]	202
OV-210	146	238	358	468	310
OV-275	629	872[a]	763	1106[b]	849
XE-60	204	381	340	493	367
Carbowax 20M	322	536	368	572	510
Diethylene glycol succinate (DEGS)	492	733	581	833	791

Note: Adapted from References 6 and 7.
[a] *n*-Butanol instead of ethanol
[b] Nitropropane instead of nitromethane

Benzene is used to assess dispersion forces and π bonding, whereas ethanol measures hydrogen bonding (as an acceptor and a donor). Each solute is run on squalane, and the liquid phase of interest at 100 °C and 20% liquid loading. The Kovats index is determined for each solute, and the difference (ΔI) between the two values on the two liquid phases is obtained. The ΔI for all the solutes is determined by the following equation:

$$\Delta I = ax + by = cz + du + es$$

where a, b, c, d, and e symbolize the five solutes listed above and x, y, z, u, and s equal the five constants that characterize the liquid phase. Since a represents benzene, x is the measure of dispersion and π-bonding forces of the liquid phases and so on. It is important to note that a has a value of 100 for benzene and a value of zero for all other solutes. Similarly, b is 100 for ethanol and zero for all others. The same also holds for the other solutes.

In 1970, McReynolds published similar constants for over 200 liquid phases;[7] however, he made two revisions in his selection of standard solutes: he used butanol instead of ethanol and expanded the list to ten compounds. Furthermore, he did not divide his values by 100. McReynolds has provided a large list of Rohrschneider's type of constants; some of these are given in Table 5. Only the first five values for benzene, *n*-butanol, 2-pentanone, nitropropane, and pyridine are given for each liquid phase, as they are considered to be the most important of the ten values.

The principal advantage of using these approaches is that selection of the best liquid phase for a given separation can be based on simple calculations.

REFERENCES

1. Ramsey, W. *Proc. Roy. Soc.* A76:111, 1905.
2. James, A. T. and Martin, A. J. P. *Biochemistry J.* 50:679, 1952.
3. Martin, A. J. P. and Synge, B. L. M. *Biochem. J.* 35:1358, 1941.
4. McNair, H. M. and Miller, J. M. *Basic Gas Chromatography*, Wiley, New York, NY, 1997, and the earlier version by McNair and Bonelli.
5. Miller, J. M. *Chromatography: Concepts and Contrasts*, Wiley, New York, NY, 1988.
6. Rohrschneider, L. *J. Chromatogr.* 22:6, 1966.
7. McReynolds, W. O. *J. Chromatogr. Sci.* 8:685, 1970.

QUESTIONS FOR REVIEW

1. Define gas–liquid chromatography, and explain how it differs from gas–solid chromatography.
2. How do you select a stationary phase for a given sample?
3. Which detector offers the greatest sensitivity for the most samples, and why?
4. Discuss briefly Rohrschneider and McReynolds constants.

10 HIGH-PRESSURE LIQUID CHROMATOGRAPHY

I. EVOLUTION OF HPLC
II. ADVANTAGES OVER GAS CHROMATOGRAPHY (GC)
III. SEPARATION PROCESS
IV. RETENTION PARAMETERS IN HPLC
   A. Peak Width
V. RESOLUTION AND RETENTION TIME
   A. Strategy of Separation
   B. Improving Resolution
VI. EQUIPMENT
VII. SEPARATION MECHANISM IN HPLC
   A. Relationship of RPLC Retention to Octanol Partition
   B. Impact of Various Physicochemical Parameters on Retention
   C. Solubility and Retention in HPLC
VIII. STATIONARY PHASE EFFECTS
   A. Adsorption/Normal Phase
   B. Reversed Phase
IX. A CASE STUDY OF RETENTION MECHANISM INVESTIGATIONS
X. MOLECULAR PROBES/RETENTION INDEX
XI. MOBILE PHASE SELECTION AND OPTIMIZATION
   A. General Considerations
   B. Modes of Chromatography
   REFERENCES
   QUESTIONS FOR REVIEW

Chromatography is growing at an enormous rate since its discovery in the early 1900s, both in terms of the number of scientific publications as well as the number of scientists who utilize it. This is largely due to the introduction of a relatively new technique called high-pressure liquid chromatography (HPLC) that offers major improvements over the old column chromatography. As a result, the total number of chromatographic applications have been increasing steadily. HPLC also provides some advantages over more recently introduced techniques such as supercritical fluid chromatography, capillary electrophoresis, and electrokinetic chromatography discussed in Chapter 11. The demand and growth of these new chromatographic techniques arise from the need to analyze a variety of complex samples that have to be analyzed to solve numerous scientific problems. In addition, this demand is driven by the perpetual need to improve the speed of analysis to save time and money.

The term liquid chromatography (LC) is applied to any chromatographic procedure in which the moving phase is liquid, as opposed to gas chromatography where a gas is utilized as a mobile phase (see discussion in Chapter 9). In classical column chromatography (see Chapter 1 as well as Section I below), gravity is the main driving force for moving the liquid mobile phase down the column. Classical column chromatography, paper chromatography (see Chapter 7), thin-layer chromatography (see Chapter 8), and HPLC are all examples of liquid chromatography.

The differences between HPLC and the old column chromatographic procedures relate to improvement in the equipment, materials used for separation, technique, and application of the theory. HPLC offers major advantages in convenience, accuracy, speed, and the ability to carry out difficult separations. Comparisons to and relative advantages of HPLC over classical LC and GC are given below (see Sections I and II) to provide the reader a better appreciation of this technique. The reader is also encouraged to look up an excellent text on HPLC by Snyder and Kirkland.[1] A short review of this technique is given here.

I. EVOLUTION OF HPLC

A review of the origination of HPLC reveals that the introduction of partition and paper chromatography in the 1940s was followed by development of gas and thin-layer chromatography in the 1950s, and the early 1960s saw the introduction of various gel or size-exclusion methods. Shortly thereafter, the need for better resolution and high-speed analyses of nonvolatile samples led to the development of HPLC.

To better understand the evolution of HPLC, let us briefly review classical column chromatography. This technique is also called packed column or open-bed chromatography. In packed column chromatography, the column is gravity fed with the sample or mobile phase; the column is generally discarded after it has been used only once. Therefore, the packing process of a column has to be repeated for each separation, and this represents significant expense in terms of both time and material. The reproducibility can vary significantly from column to column. Sample application, if done correctly, requires some skill and time on the part of the operator. Solvent flow is dependent upon the gravity flow of the solvent through the column, and individual fractions are collected manually. Typical separation requires several hours. Detection and quantification are achieved by the analysis of each fraction. Normally, many fractions are collected, and their processing requires significant time and effort. The results are recorded as a bar graph of sample concentration versus fraction number.

High Pressure or High Performance. Frequently this technique (HPLC), which utilizes high pressure for pumping liquids and uses more efficient columns, is simply called LC. Unfortunately, the terms LC and HPLC have been used interchangeably by various researchers; however, it should be recognized that the appropriate term is HPLC and should be used preferentially. At times, this technique is called high-performance liquid chromatography (which has the same abbreviation, viz., HPLC) because it does provide high performance as compared with classical liquid chromatography. The use of this term must be made judiciously since all applications of HPLC are not necessarily based on high performance.

HPLC employs reusable columns that have a frit at each end to contain the packing, so that numerous individual separations can be carried out on a given column. Since the cost of an individual column can be distributed over a large number of samples, it is possible to use more expensive column packings for obtaining high performance and also to spend more money on commercially packed columns. Alternatively, more time can be spent on the careful packing of a column for best results. Precise sample injection is achieved easily and rapidly in HPLC by using either syringe injection or a sample valve. High-pressure pumps are used to obtain the desired solvent flow. This has a decided advantage in that controlled, rapid flow of the solvent is achieved through relatively impermeable columns. The controlled flow results in more reproducible operation, which results in greater accuracy and precision in HPLC analysis. High-pressure operation leads to better and faster separations. Detection and quantification in HPLC are achieved with continuously monitoring detectors of various types. These yield a final chromatogram without intervention by the operator. The end result is an accurate record of the separation with minimum effort.

A large number of separations can be performed by HPLC by simply injecting various samples and then applying appropriate final data reduction, although the column and/or solvent may require a change for each new application. It should be obvious, based on these comments, that HPLC is considerably more convenient and less operator-dependent than classical column chromatography. The greater reproducibility and continuous quantitative detection in HPLC allows more reliable qualitative analysis or more accurate and precise quantitative analysis than classical column chromatography.

II. ADVANTAGES OVER GAS CHROMATOGRAPHY (GC)

Compared with older chromatographic methods, gas chromatography (see Chapter 9) provides separations that are faster and better in terms of resolution. It can be used to analyze a variety of samples. However, GC simply cannot handle many samples without derivatization, because the samples are not volatile enough and cannot move through the column or because they are thermally unstable and decompose under the conditions of separations. According to estimates, GC can sufficiently separate only 20% of known organic compounds without prior chemical alteration of the sample.

An important advantage of HPLC over GC is that it is not restricted by sample volatility or thermal stability. It is also ideally suitable for the separation of macromolecules and ionic species of biomedical interest, labile natural products, and diverse less stable and/or high molecular weight compounds, such as the following:

- Amino acids
- Drug products
- Dyes
- Explosives
- Nucleic acids
- Pharmaceutical active ingredients and impurities
- Pharmaceutical metabolites
- Plant active ingredients
- Plant pigments
- Polar lipids
- Polysaccharides
- Proteins and peptides
- Recombinant products
- Surfactants
- Synthetic polymers

HPLC offers a number of other advantages over GC. The majority of difficult separations are often more readily attained by HPLC than by gas chromatography because of the following reasons:

- Since both phases used in HPLC participate in the chromatographic process, as opposed to only one in GC, more selective interactions are possible with the sample molecule.
- A large variety of unique column packings (stationary phases) provide a wide range of selectivity.
- Lower separation temperatures are frequently used in HPLC.

Chromatographic separation results from the particular interactions between sample molecules and the stationary and moving phases. These interactions are inherently absent in the moving gas phase of GC, but they do exist in the mobile liquid phase of HPLC. As a result, they provide an additional variable for regulating and improving separation. The great variety of fundamentally distinct stationary phases has been found useful in HPLC, again allowing a more extensive variation of these selective interactions and greater possibilities for separation. The chromatographic separation is improved in general as the temperature is lowered, because intermolecular interactions become more effective. This works favorably for procedures such as HPLC that are usually carried out at ambient temperature. HPLC also offers a number of unique detectors that have found limited application in GC:

- Visible wavelength detectors such as those that can be used for measurement of color of sample components by colorimeteric reactions
- Electrochemical detectors
- Refractive index detectors
- UV detectors
- Fluorescent detectors
- NMR as an on-line detector
- Mass spectrometric detectors

10 HIGH-PRESSURE LIQUID CHROMATOGRAPHY

Even though detectors used for GC are generally more sensitive and provide particular selectivity for many types of samples, the available HPLC detectors offer unique advantages in a variety of applications. In short, it is a good idea to recognize the fact that HPLC detectors are favored for some samples, whereas GC detectors are better for others.

Another significant advantage of HPLC over gas chromatography is the relative ease of sample recovery. In HPLC, simply placing an open vessel at the end of the column allows separated fractions to be collected with ease. Recovery is quantitative, and separated sample components are easily isolated for identification by supplementary techniques. The recovery of separated sample bands in GC is also feasible, but is generally less convenient.

III. SEPARATION PROCESS

In HPLC, the mobile phase is constantly passing through the column at a finite rate. Sample injections are made rapidly in the dynamic state in the form of a narrow band or plug (see Fig. 1), which provides a significant advantage over older stop-flow injection techniques in HPLC. After a sample has been injected as a narrow band, the separation can then be envisioned as a three-step process as shown in the figure:

- Step 1: As the sample band starts to flow through the column, a partial separation of components X, Y, and Z components begins.

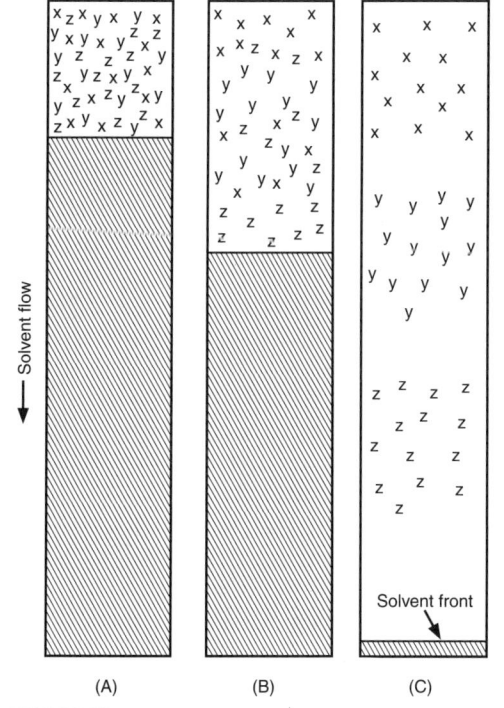

FIGURE 1 Diagrammatic representation of separation.

- Step 2: The separation improves as the sample moves further through the column.
- Step 3: The three compounds are essentially separated from each other.

At step 3, we can see two characteristic features of chromatographic separation:

1. There are different migration rates of various compounds (solutes) in the sample.
2. Each solute's molecules are spread along the column.

Differential migration in HPLC relates to the varying rates of movement of different compounds through a column. For example, in Figure 1 compound Z moves most rapidly and leaves the column first; compound X moves the slowest and leaves the column last. As a result, compounds X and Z are gradually separated as they move through the column.

The differential migration in HPLC is also related to the equilibrium distribution of different compounds such as X, Y, and Z between the stationary phase and the flowing solvent(s), or mobile phase. Since sample molecules do not progress through the column while they are in the stationary phase, the speed with which each compound travels through the column (u_x) is determined by the number of molecules of that compound in the moving phase, at any moment. As a result, compounds such as X, whose molecules spend most of the time in the stationary phase, travel through the column slowly. Compounds such as Z, whose molecules are in the mobile phase most of the time, progress through the column more rapidly. The molecules of the solvent or mobile phase travel at the fastest possible rate, except in size-exclusion chromatography where molecular size is used to effect the separation.

Another characteristic of chromatographic separation is spreading of molecules such as X along the column. The molecules start out as a narrow band (see Fig. 2A) that gradually broadens as it moves through the column.

As mentioned in Chapter 9, the differences in molecular migration occur as a result of physical or rate processes. Some of these processes are summarized below.

Eddy Diffusion. One of the processes leading to molecular spreading is caused by multiple flow paths and is called eddy diffusion (see Fig. 2B). Eddy diffusion results from distinct microscopic flow streams that the solvent tracks between different particles within the column. Sample molecules, consequently, move on different paths through the packed bed, depending on which flow streams they follow. It can be readily visualized that some of these paths are quite slow and others are very fast.

Mass Transfer. Another contribution to molecular spreading is seen in Figure 2C: mobile phase mass transfer. This pertains to differing flow rates for different regions of a single flow stream or path between neighboring particles. It can be seen in Figure 2C, where the flow stream between particles 1 and 2 is shown, that liquid adjacent to a particle would move very slowly, whereas liquid in the middle of the flow stream would move most rapidly. Thus, at any given moment, sample molecules close to the particle move a short distance

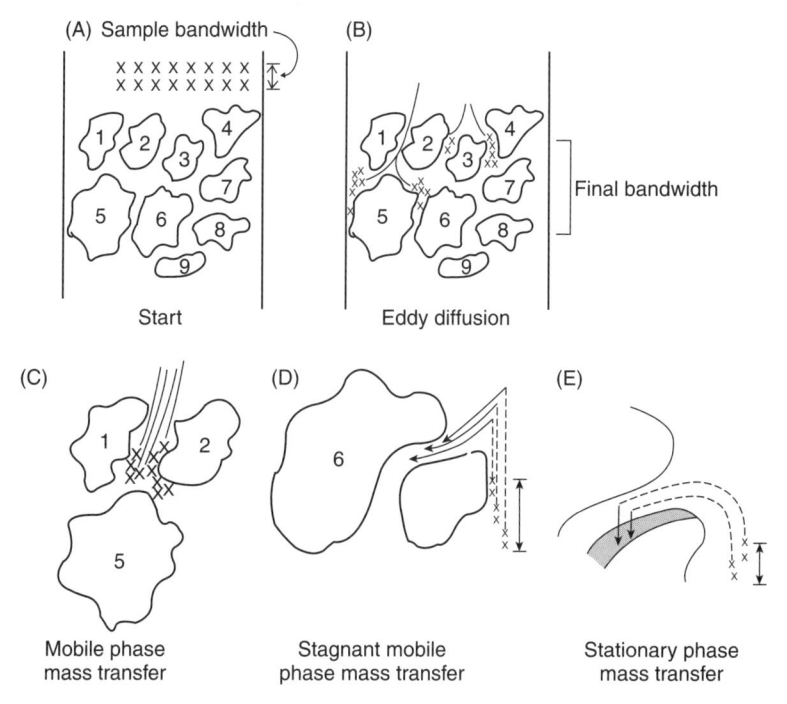

FIGURE 2 Contributions to band spreading.

and sample molecules in the middle of the flow stream travel a greater distance. Spreading of molecules along the column is therefore increased.

Figure 2D demonstrates the effect of stagnant mobile-phase mass transfer on molecular spreading. When porous column-packing particles are used, the mobile phase within the pores of the particle is stagnant, or immobile. (In Figure 2D one such pore is shown for particle 6.) Sample molecules move into and out of these pores by diffusion. Those molecules that diffuse a short distance into the pore and then diffuse out, go back to the mobile phase rapidly, and move a certain distance down the column. Molecules that diffuse further into the pore spend more time there and less time in the external mobile phase. Consequently, these molecules travel less distance down the column, and molecular spreading is further increased.

Figure 2E shows the result of stationary-phase mass transfer. After molecules of sample diffuse into a pore, they move into the stationary phase (shaded region) or become bound to it in some manner. Molecules return to the mobile phase sooner if they spend only a short time moving into and out of the stationary phase; consequently, they move further down the column.

Longitudinal Diffusion. The longitudinal diffusion, provides an additional contribution to molecular spreading. Regardless of whether the mobile phase within the column is moving or resting, sample molecules are inclined to diffuse haphazardly in every direction. Over and above the other effects shown in Figure 2, this contributes to further spreading of sample molecules along the column. Longitudinal diffusion is usually an insignificant effect, but is important at low mobile-phase flow rates for small-particle columns.

We can then eliminate L between these last two equations and get

$$t_R = \frac{ut_0}{u_x} \qquad (8)$$

Relating Retention Time to Capacity Factor. The following equation can then be derived by substituting for the value of u_x from Eq. (5):

$$t_R = t_0(1 + k') \qquad (9)$$

This expresses t_R as a function of the fundamental column parameters t_0 and k'; t_R can vary between t_0 (for $k' = 0$) and any larger value (for $k' > 0$). Because t_0 varies inversely with u (solvent velocity), t_R also does so. For a given column, mobile phase, temperature, and sample component X, k' is normally constant for sufficiently small samples. As a result, t_R value for a given compound X is defined by the chromatographic system. Therefore, t_R value of a compound can be used to identify it tentatively by comparison with a t_R value of a known compound.

On rearrangement, Eq. (9) gives an expression for k':

$$k' = \frac{t_R - t_0}{t_0} \qquad (10)$$

It is generally necessary to determine k' for one or more bands in the chromatogram to plan a strategy for improving separation. Equation (10) gives a simple, rapid basis for estimating k' values in these situations; k' equals the distance between t_0 and the band center, divided by the distance from injection to t_0.

The meaningful column parameter t_0 may be determined in a number of ways. Usually, the center of the first band or baseline disturbance following sample injection denotes t_0. If there is any doubt about the position of t_0, a weaker solvent (or other unretained compound) may be injected as a sample, and its t_R value will equal t_0.

A weaker solvent provides greater k' values and stronger sample retention than the solvent used as the mobile phase (see Table 1 for a list of solvents according to strength). For example, hexane could be injected as a sample to determine t_0 if chloroform was to be used as the mobile phase in liquid–solid chromatography. When the flow rate of the mobile phase through the column remains constant, t_0 is the same for any mobile phase. If flow rate fluctuates by some factor x, t_0 will change by the factor $1/x$.

Retention Volume. From time to time, retention in HPLC is measured in volume units (ml) rather than in time units (s). Hence, the retention volume V_R is the total volume of mobile phase required to elute the center of a given band X, that is, the total solvent flow in time between sample injection and appearance of the band center at the detector. The retention volume V_R is equal to retention time t_R multiplied by the volumetric flow rate F (ml/min) of the mobile phase through the column:

$$V_R = t_R F \qquad (11)$$

TABLE I Solvent Strength of Some Useful Solvents for HPLC

Solvent	Silica	Alumina	Selectivity group*
n-Hexane	0.01	0.01	–
Chloroform	0.26	0.40	VIII
Methylene chloride	0.32	0.42	V
Ethyl acetate	0.38	0.58	VI
Tetrahydrofuran	0.44	0.57	III
Acetonitrile	0.50	0.65	VI
Methanol	0.7	0.95	II

*See Section XI for selectivity groups classification.

The total volume of solvent within the column (V_M) is

$$V_M = t_0 F \qquad (12)$$

Substituting for F gives

$$V_R = V_M(1 + k') \qquad (13)$$

A. Peak Width

The peak width relates to retention time of the peak and the efficiency of a chromatographic column, or total number of theoretical plates (N) offered by it.

Total Number of Theoretical Plates. Mathematically, the width of a chromatographic peak (t_w) can be expressed as a number of theoretical plates (N) of the column:

$$N = \frac{16(t_R)^2}{t_w} \qquad (14)$$

The quantity N is approximately constant for different bands or peaks in a chromatogram for a given set of operating conditions (a particular column and mobile phase, with fixed mobile phase velocity and temperature). Therefore, N is a useful measure of column efficiency: the relative ability of a given column to provide narrow bands (small values of t_w) and improved separations.

Equation (14) predicts that peak width will increase proportionately with t_R because N remains constant for different bands in a chromatogram, and this is usually found to be true. Insignificant exceptions to the constancy of N for different bands do occur,[1] and in gradient elution chromatography all bands in the chromatograms have a tendency to be of equal width.

Since HPLC peaks broaden as retention time increases, later-eluting bands exhibit a corresponding reduction in peak height and gradually vanish into the baseline. The quantity N is proportional to column length L, so that if other factors are equal, an increase in L results in an increase in N and better separation.

Height Equivalent of a Theoretical Plate. The proportionality of N and L can be expressed in terms of the following equation:

$$N = L/H \tag{15}$$

where H = height equivalent of a theoretical plate (HETP), N = total number of theoretical plates, and L = length of the column.

The quantity H (equal to L/N) measures the efficiency of a given column (operated under particular operating settings) per unit length of the column. Small H values indicate a more efficient columns and large N values. A very important objective in HPLC is to arrive at small H values that lead to maximum N and highest column efficiencies.

Based on various theoretical and practical observations, we can draw a number of conclusions. The value of H decreases with:

- small particles of column packing
- low mobile-phase flow rates
- less viscous mobile phases
- separations at high temperatures
- small sample molecules

V. RESOLUTION AND RETENTION TIME

A simple and yet very important goal in HPLC is to obtain satisfactory separation of a sample mixture. To achieve this goal, we need to have some quantitative measure of the relative separation or resolution attained. The resolution, R_s, of two adjacent peaks 1 and 2 is defined as equal to the distance between the centers of two peaks, divided by average peak width (see Fig. 3):

$$R_s = \frac{(t_2 - t_1)}{1/2(t_{w1} + t_{w2})} \tag{16}$$

where t_1 and t_2 = the retention times of peaks 1 and 2 and, t_{w1} and t_{w2} = their peak width values

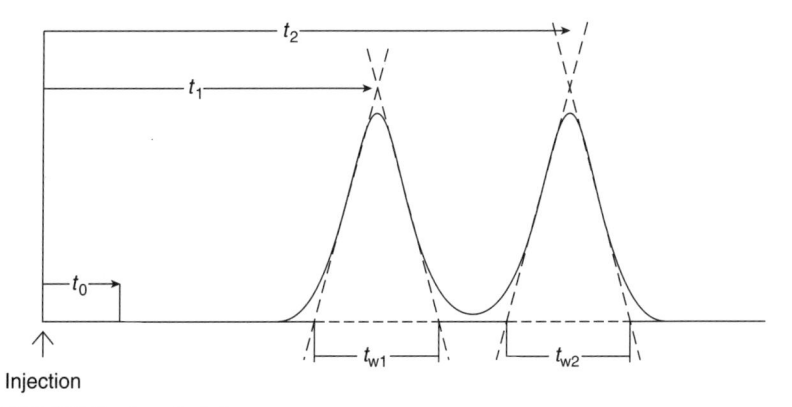

FIGURE 3 A typical chromatogram.

When $R_s = 1$, as in Figure 3, the two peaks are reasonably well separated; that is, only 2% of one peak overlaps the other. Larger values of R_s, mean better separation, and smaller values of R_s indicate poorer separation. For a given value of R_s, peak overlap is a greater problem when there is a large difference in the peak sizes.

The resolution value, R_s, of Eq. (16) helps to describe a given separation. To regulate separation or resolution, we must be aware of how R_s varies with experimental parameters such as k' and N. We can derive a simplified relationship for two closely spaced (i.e., overlapping) peaks (1). Based on Eq. (9), $t_1 = t_0(1 + k_1)$ and $t_2 = t_0(1 + k_2)$, where k_1, and k_2 are the k' values of bands 1 and 2.

Since t_1 is approximately equal to t_2, t_{w1} will be approximately equal to t_{w2} based on Eq. (14) if we assume N is constant for both bands. Then Eq. (16) can be written as follows:

$$R_s = \frac{t_0(t_2 - t_1)}{t_{w1}} \qquad (17)$$

Equation (14) similarly gives $t_{w1} = 4t_1/N^{1/2} = 4t_0(1 + k_1)/N^{1/2}$. Substituting this value in Eq. (17) gives

$$R_s = \frac{(k_2 - k_1)N^{1/2}}{4(1 + k_1)}$$
$$= \frac{1(k_2 - 1)}{4k_1} \frac{N^{1/2}(k_1)}{1 + k_1} \qquad (18)$$

Separation Factor and Resolution. The separation factor α is equal to k_2/k_1. By inserting α and an average value of k' for k_2 and k_1, we can simplify the resolution equation as follows:

$$R_s = 1/4(\alpha - 1)N^{1/2}(k'/1 + k) \qquad (19)$$

It is important to recognize that a number of assumptions were made in deriving Eq. (17) that led us to the simplified Eq. (19). A more fundamental form of resolution expression is given in Eq. (20). To get a more accurate equation, the actual values of the peak widths and their respective capacity factors should be used; however, for most practical purposes the above equation or its original form, Eq. (16), are satisfactory.

A. Strategy of Separation

A strategy to design a successful HPLC separation should involve the following steps:

1. Select the method most suitable for your sample, based on solubility of sample and other relevant physical properties. For example, the preliminary choice of a method would include one of the following methods:
   - normal phase
   - reversed phase

- ion exchange
- size exclusion

2. Choose a suitable column based on this method selection. For example, in reversed-phase HPLC, the choice may simply entail selecting C-8 or C-18.
3. Decide on a simple mobile phase for initial evaluations. For example, methanol/water (50:50) may be a good starting point in the case of RPLC.
4. Make desirable changes in the mobile phase in terms of proportions of solvent or vary solvents or include suitable additives.
5. Utilize specialized techniques such as gradient elution or derivatization to enhance detectability and/or improve separations.

B. Improving Resolution

Resolution can be improved by varying the three terms α, N, or k' in the resolution equation:

$$R_s = 1/4(\alpha - 1/\alpha)N^{1/2}(k'_2/1 + k'_2) \qquad (20)$$

The effect of changes in k', N, or α on sample resolution is shown in Figure 4. For example, increased separation factor α is followed by a displacement of one band center relative to the other and a rapid enlargement of R_s. The peak height or retention time is not significantly affected for a moderate change in α. The increase in plate number, N, produces a narrowing of the two bands and increased band height. The retention time is not affected if there is no change in the weight ratio of the sample.

The most significant effect on separation is caused by change in k' values. If k' for the initial separation occurs within the range $0.5 < k' < 2$, a decrease in k' can make a separation look bad. On the other hand, an increase in k' can provide a significant increase in resolution. Nonetheless, as k' is increased, band heights quickly decrease and separation time becomes greater.

When it is necessary to improve R_s, k' should be increased into the optimum range, which is $1 < k' < 10$. When k' is less than 1, R_s increases rapidly with increase in k'. For values of k' greater than 5, R_s increases very little with further increase in k' (Table 2).

If k' were infinity, its proportionality would be 1.00. This means that there is no major improvement over that seen with k' of 5, as shown in the table. These observations suggest that the optimum k' is in the range of 1 to 10. It is important to note here that no other modification in separation condition affords as great an improvement in R_s value for such little effort.

Solvent strength is a good method of controlling k' values in HPLC. When it is necessary to increase the k' value, a weaker solvent should be used. For example, in reversed-phase separations, solvent strength is greater for pure methanol than for pure water. The right proportionality of these solvents has to be found to obtain the optimum separation.

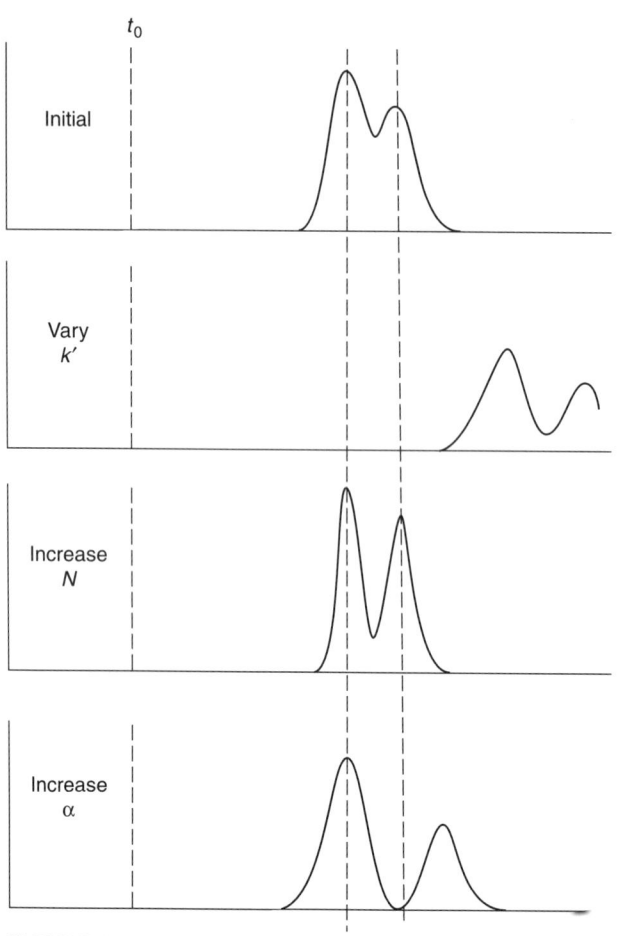

FIGURE 4 Effect of α, k', or N on resolution.

TABLE 2 Impact of k' on R_s Value*

k'	$k'/1+k'$
0	0
1	0.5
2	0.67
5	0.83
10	0.91

*See Eq. (20).

When k' is already within the optimum range of values and resolution is still marginal, the best solution is to increase N. This is generally accomplished by increasing the column length or by decreasing the flow rate of the mobile phase.

When R_s is still small even though k' is optimum, an increase in N would not be very helpful because it creates an unusually prolonged separation time.

In this situation, increasing α value would be more desirable. However, it should be recognized that predicting the right conditions for the necessary change in α is not a simple procedure (for more detailed discussion, see Section VII).

VI. EQUIPMENT

A photograph of an HPLC instrument is shown in Figure 5 and the schematic diagram of an HPLC instrument is given in Figure 6. The equipment has to be designed and produced to assure correct volumetric delivery of the mobile phase and injected sample, and low-noise detectors must be utilized so that low concentrations of samples can be analyzed. Some of the important criteria for instrumentation are summarized in Table 3.

A comparative evaluation of pumps used in HPLC is given in Table 4. The pumps specifically designed for HPLC are able to provide constant flow of the mobile phase against column pressure up to 10,000 psi. However, it is important to recognize that most HPLC separations are run at pressures less than 6000 psi, and the advantage of pumps with a high-pressure rating is marginal.

Most chromatographers limit themselves to binary or tertiary gradient systems; however, it should be noted that pumps capable of quaternary gradients are available and should be considered in the equipment selection process to allow greater versatility in method development.

As mentioned above, a number of detectors can be used with HPLC. The most useful detector for characterization of separated compounds is MS. A photograph of an HPLC/MS is shown in Figure 7.

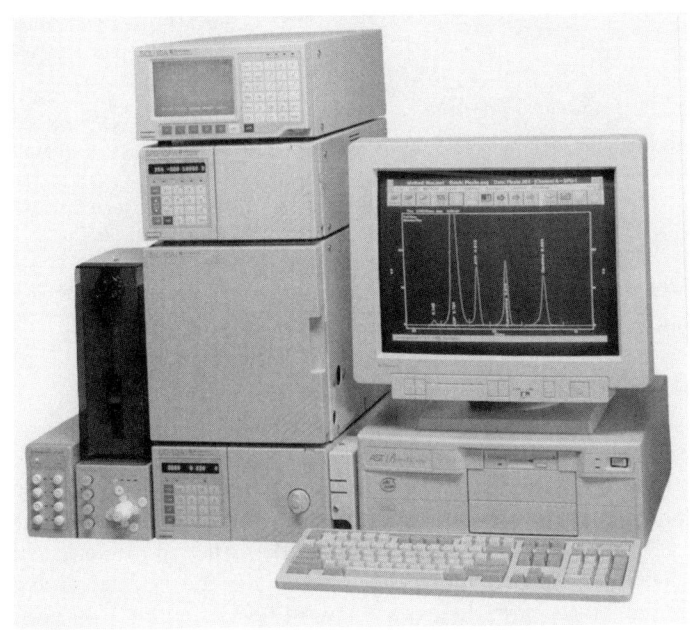

FIGURE 5 A photograph of an HPLC instrument (courtesy of Shimadzu Instruments).

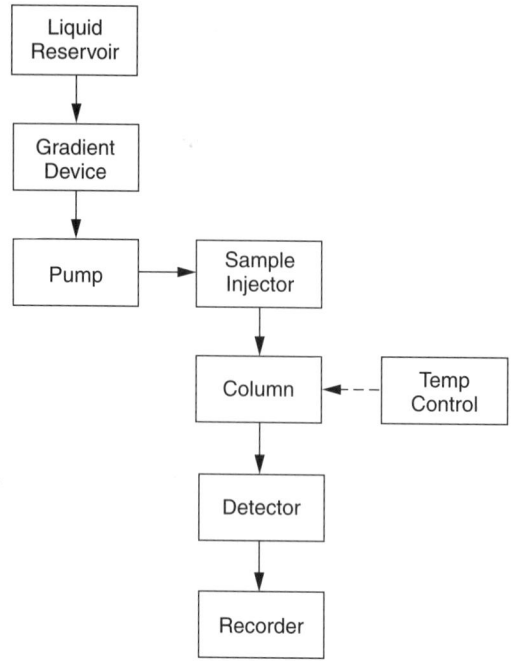

FIGURE 6 A schematic of HPLC equipment.

TABLE 3 Instrument Requirements for HPLC

Criteria	Characteristics	Instrument requirements
Reproducibility	Control various operational parameters	Control precisely: mobile phase composition temperature flow rate detector response
Detectability	High detector response Narrow peaks	Good signal-to-noise ratio Efficient columns
Sample variety	Useful for a variety of samples	Variety of detectors and stationary phases
Speed	Selective and efficient columns High flow rate Fast data output	Low dead-volume fittings High-pressure pumps Fast-response recorders and automatic data handling

VII. SEPARATION MECHANISM IN HPLC

As mentioned before, selectivity in HPLC involves both the mobile phase and the stationary phase. This distinguishes it from GLC where the stationary phase is primarily responsible for the observed separation.

It is also important to recognize that the stationary phase can be influenced by its environment. The reason for the erroneous assumption to the contrary arises from the experiences in GLC where the stationary phase is minimally influenced by the gaseous mobile phase. In the early 1960s,

10 HIGH-PRESSURE LIQUID CHROMATOGRAPHY

TABLE 4 Evaluation of Various HPLC Pumps

	Reciprocating						Positive displacement		Pneumatic		
Pump characteristic	Simple single-head	Single-head, smooth pulse	Simple dual-head	Dual-head, compressibility corrected, smooth pulse	Dual-head, closed loop flow control	Triple head low-volume	Syringe-type	Hydraulic amplifier	Simple	Amplifier	Amplifier with flow control
Resettable	+	+	++	++	++	++	++	++	−	−	+
Drift	+	+	+	++	++	++	++	+	−	+	+
Low Noise	−	+	+	++	++	++	++	++	+	+	++
Accurate	+	+	+	+	++	++	+	+	−	−	+
Versatile	−	++	++	++	++	++	−	+	−	++	+
Low Service	+	+	+	+	+	+	−	+	++	+	−
Durable	+	++	+	+	+	+	+	−	++	++	++
Cost	Low	Moderate	Moderate	High	Very high	Very high	Moderate to high	Moderate	Low	Moderate	High
Consistent flow	Yes	Yes	Yes	Yes	Yes	Yes	Yes	Yes	No	No	No
Constant pressure	No	No	No	Yes	No	No	No	Yes	Yes	Yes	Yes

Note: Adapted from Reference 1.
++ = optimum, + = satisfactory, − = needs improvement.

FIGURE 7 A photograph of an HPLC/MS instrument (courtesy of Waters Instruments).

it was demonstrated by Ahuja et al.[2] that even in GLC, the stationary phase could be affected by the composition of the mobile phase. Compounds such as water, when present in the carrier gas, can be retained by the stationary phase and then participate in the partition process when a sample is injected. It was shown that the stationary phase is a dynamic system that is liable to shift with the composition of the mobile phase. This is an even more important consideration in HPLC, where multiple components are constantly bathing the stationary phase. Prior to developing an understanding of the separation mechanism in HPLC, it would be a good idea to review the physicochemical basis of retention in HPLC (for more details, see Chapter 2 in Reference 3).

Since selectivity in HPLC involves both the stationary and mobile phases,[3-9] it is important to note that the solvent strength of the mobile phase, as compared to the stationary phase (composed of mobile-phase components reversibly retained by the bonded phase and silica support), determines the elution order or k' of the retained components. Unfortunately, columns with the same stationary phase may exhibit significant variabilities from one manufacturer to another, and even from the same manufacturer.[3-5] Based on discussions at various scientific meetings, it would seem that this situation has not changed much in recent years. Variabilities can occur in the packing process even where all other conditions are supposedly constant. These factors have to be considered in understanding how separations occur in HPLC.

The technique used for selecting a particular chromatographic separation generally involves maximizing the column efficiency while choosing a stationary and/or mobile phase that is thought to be suitable for the sample at hand. There is a real lack of quantitative guidelines for optimizing selectivity decisions made pertaining to appropriate experimental conditions, and the techniques to achieve this end have generally been based primarily on the experience of the analyst as guided by literature.[6] An alternative approach to optimizing the system selectivity, one more rational than that of trial-and-error, would be understanding the physicochemical basis of separations, followed by modeling of the system at hand, and then optimizing the system and conditions within the boundaries of the required resolution. Several theories have been proposed to explain physicochemical interactions that occur between the solute and the mobile and stationary phases as well as those that may occur between the mobile and stationary phases themselves.[6,7]

It is generally recognized that in spite of the enormous popularity of reversed-phase high-performance liquid chromatography (RPLC), which may be broadly classified as involving distribution of nonpolar or moderately polar solute between a polar mobile phase and a relatively nonpolar stationary phase, details of solute retention and selectivity are not fully understood. Such an understanding is critical for informed control and manipulation of separations that ultimately require an adequately detailed description of how retention and selectivity depend on the mobile- and stationary-phase variables.[7] The character of the solute distribution process in RPLC, that is, the retention mechanism, has also been a topic of much study, discussion, and speculation. The most meticulous treatment to date is based on the solvophobic theory, with principal focus on the role of the mobile phase. Invoking "solvophobic" interactions, that is, exclusion of the less-polar solute molecule from the polar mobile phase, with the later sorption by the nonpolar stationary phase shapes solute distribution. This implies that the mobile phase "drives" the solute toward the stationary phase in lieu of the solute's being driven by any inherently strong attraction between the solute and the stationary phase. The fundamental premise of the theory is reasonable, and agreement with experimental data is generally good; nonetheless, the description is incomplete in that it does not give a sufficiently detailed explanation of the dependence of solute retention and selectivity on the stationary-phase variables. Furthermore, it has been reported that, under particular chromatographic conditions (very polar mobile-phase solvent and relatively long n-alkyl chains), solute distribution in RPLC seems to approach that of partitioning between two bulk liquid phases, suggesting the liquidlike behavior of unswollen n-alkyl chains.

There is little published data providing systematic studies of retention volumes as a function of composition of the eluent over the whole composition range of binary solvents. To correct this situation, a general equation for HPLC binary solvent retention behavior has been suggested[8] that should help produce a chromatographic retention model to fit Eq. (21):

$$\frac{1}{V_{AB}} = \varphi A \frac{1}{V_A} + \frac{b\varphi_B}{b'\varphi_B} + \frac{\varphi_B}{V_B} \qquad (21)$$

where V_{AB} = the corrected elution volume of the solute (experimental elution volume-void volume) in the binary eluent A + B, with volume fractions φA and φB, respectively; V_A = the corrected elution volume for the solute with pure A as eluent; V_B = the corrected elution volume for the solute with pure B as eluent; and b and b' = constants (the term that includes them is reminiscent of the Langmuir adsorption isotherm).

A. Relationship of RPLC Retention to Octanol Partition

Although the mechanism responsible for retention in RPLC on alkyl-bonded stationary phases is still not fully understood, it is known that the RPLC retention parameters are often strongly correlated to the analyte's distribution coefficient in organic solvent/water. In general, the relationship between liquid/liquid (LL) distribution and RPLC retention are of the form of the dimensionless Collander-type equations, for example, see Eq. (22) below.

$$\log k_d = a_1 + b_1 \log k' \qquad (22)$$

where K_d = the solute's organic solvent/water distribution coefficient, k' = chromatographic capacity ratio ($k' = t_r - t_0/t_0$), t_r and t_0 = solute retention time and mobile phase "hold-up" time, respectively, and a and b = coefficients whose magnitudes depend on the LL distribution and RPLC systems.

The general validity of this type of relationships depends on:

- The physicochemical state (temperature, degree of ionization) of solutes in both liquid/liquid distribution and RPLC
- The RPLC system used, with emphasis on the mobile phase, principally on the type, but also to some degree on the concentration of organic cosolvents (modifiers) used
- The type of organic solvent used in the distribution system modeling an RPLC system

It has been shown that RPLC capacity factors of unionized solutes realized using aqueous methanol mobile phases can ordinarily be correlated with octanol/water distribution coefficients for neutral, acidic, and basic compounds, while no such overall correlation can be made with alkane/water distribution parameters.[10,11] HPLC can be utilized for the measurement of octanol/water partition coefficients even when only a small amount of sample is available. Solutes are equilibrated between the two phases by the conventional shake-flask method, which entails shaking them together and then analyzing by HPLC sampled portions of both phases.[12] The results thus obtained with a variety of substances having partition coefficients in the range of 10^{-3} to 10^4 exhibited excellent agreement with literature data. The method is rapid and has the advantages that small samples are sufficient, the substances do not need to be pure, and it is not necessary to know the exact volume of the phases. Furthermore, it could easily be developed into a micromethod for measurements with submicrogram quantities. Table 5 shows a comparison of data obtained by HPLC with the classical method.

Data collected with a simple RPLC procedure have been found to be in good conformity with 1-octanol shake-flask partition or distribution coefficients over a 3.5 log range.[13] A chemically bonded octadecylsilane support is coated with 1-octanol. Using 1-octanol-saturated buffers as mobile phases, a stable baseline (compared to 1-octanol coated on silica) is quickly achieved, and the log relative retention times are well correlated with unit slope to log distribution or partition coefficients obtained from the classical

TABLE 5 Comparison of Log P Values by HPLC with Literature Values

Name	HPLC	Literature
tert-Butylbenzene	4.07	4.11
Propylbenzene	3.44	3.62
Toluene	2.68	2.58
Acetophenone	1.58	1.66
Resorcinol	0.88	0.80
Hydroquinone	0.54	0.56
Caffeine	−0.05	−0.07

Note: Adapted from Reference 12.

TABLE 6 Comparison of log P and Apparent pK_a Determined by HPLC with Literature Values

	HPLC		Literature	
Compound	pK_a	log P	pK_a	log P
Naproxen	4.28	3.21	4.53	3.18
	4.21	3.20	4.39	
Benzoic acid	4.33	1.78	4.20	1.87
	4.38	1.77	4.18	
Salicylic acid	3.52	2.00	3.00	2.23
	3.29	2.18		
p-Toluic acid	4.30	2.22	4.37	2.27
	4.41	2.26		

Note: Adapted from Reference 13.

shake-flask procedures. Only relatively basic, unhindered pyridines vary, probably because of binding with residual silanol sites.

In addition, if the apparent pK_a of an ionizable compound falls within the pH operating range of the column support, the apparent pK_a usually may be determined concurrently with log P by measuring the log distribution coefficient at several pH values (Table 6). The primary advantages of the method are that it gives quick results, calls for little material, and can tolerate impurities.

Using an unmodified commercial octadecylsilane column and a mobile phase composed of methanol and an aqueous buffer, a linear relationship has been demonstrated between the literature log P values of 68 compounds and the logarithms of their k' values.[14] For the determination of the partition coefficients of unknowns, standards are used to calibrate the system, the choice being dependent on the hydrogen-bonding characteristic of the compounds being evaluated. The whole method is rapid and broadly adaptable to give log P data comparable to results achieved by classical and other correlation methods.

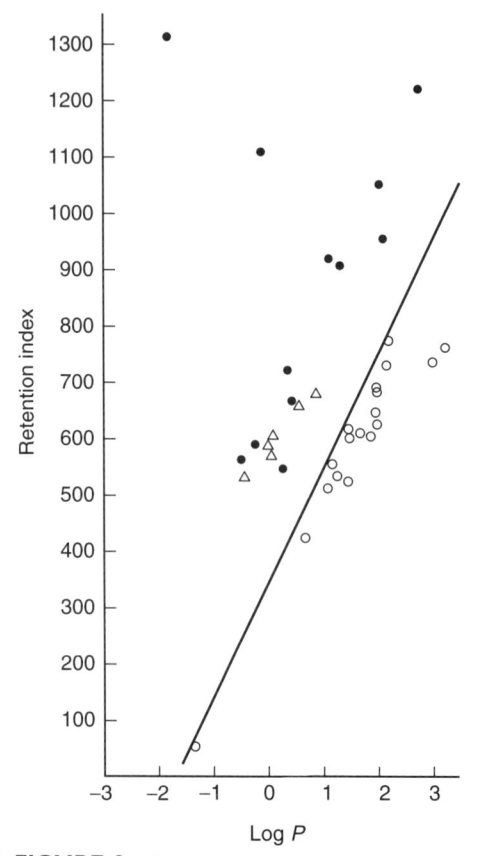

FIGURE 8 Correlation between observed retention indices and octanol partition coefficients (15).

Studies with barbiturates have revealed that the logarithm of the retention time is linearly related to the octanol/water partition coefficients.[15,16] It has been observed that the retention index of the drug is linearly related to the octanol/water partition coefficient (log P) and that results are very close to that of the 2-keto alkane standard (note the solid line in Figure 8).

The retention data of catecholamine derivatives in reversed-phase chromatography with octadecyl-silica stationary phase and aqueous eluent has been analyzed. Good agreement is observed between the observed and predicted k' values.[17] Data obtained with different C-18 stationary phases at various temperatures suggest that quantitative structure–retention relationships can be transformed from one reversed-phase system to another as long as the eluent composition is the same. Kalizan[18] observed that C-18 columns without any additional coating give results that are precise enough for quantitative structure–activity relationship (QSAR) purposes and are useful over a wide range of operational conditions.

It has been suggested by Tomlinson[19] that a "hydrophobic bond" is formed when two or more nonpolar groups in an aqueous medium come into contact, thereby decreasing the extent of interaction with the adjacent water molecules and resulting in the release of water originally bound by the molecules. The hydrophobic bond is recognized as having a complex character, involving polar and nonpolar interactions; the hydrophobic bond concept has been practical in explaining association of organic and biologic molecules in aqueous solution. In QSAR models, the capacity of a compound to partition between a relatively nonpolar solvent and water is customarily used as a measure of its hydrophobic character.

Martin was the first to suggest that a substituent changes the partition coefficient of a substance by a given factor that depends on the characteristics of the substituent and the two phases used, but not on the rest of the molecule (see Chapter 4). Martin's treatment presumes that, for any stated solvent system, the alteration in retention (ΔR_{min} TLC) caused by the introduction of group X into a parent structure is of uniform value, on condition that its substitution into the parent structure does not result in any intramolecular interactions with other functions in the structure. Conversely, it can be appreciated that if the introduction of a group into a structure brings about a breakdown in the additivity principle, then inter- or intramolecular effects are likely to be more significant within the substituted structure. These effects are as follows.

- Chain branching
- Electronic effects
- Intramolecular hydrogen bonding
- Intramolecular hydrophobic bonding
- Steric effects (including ortho effect)

Hydrophobicity and the partition coefficient may be related to the solubility of the solutes in water (also see Section VII. C). The partition coefficient (P) between octanol and water may be depicted as the pi constant of Hansch or the log P values of Rekker. The log P values determined from the fragmental constant are then used for the optimization of reversed-phase liquid chromatography. This method, however, is insufficient for developing an optimization system for the mixtures of different types of compounds.[20]

B. Impact of Various Physicochemical Parameters on Retention

Discussed here is the impact of a variety of physicochemical parameters excluding solubility (see Section VII. C) on retention in HPLC. The van der Waals volume can be connected to the hydrophobicity of the solutes, and retention of molecular compounds can be anticipated from their van der Waals volumes, pi energy, and hydrogen-bonding energy effects.[21-23] The isomeric effect of substituents cannot be predicted with good precision because this is not simply related to Hammett's σ or Taft's σ^* constants. On the contrary, the hydrophobicity is related to enthalpy.[24] Retention times of nonionizable compounds have been measured in 70% and 80% acetonitrile/water mixtures on an octadecyl-bonded silica gel at 25°C–60°C, and the enthalpy values were obtained from these measurements.

Retention volumes of monosubstituted benzenes, benzoic acid, phenols, and anilines have been measured in RPLC.[25] Buffered acetonitrile/water and tetrahydrofuran/water eluents were used with an octadecylsilica adsorbent. From the net retention volumes, a substituent interaction effect was calculated and described with the linear free-energy relationship developed by Taft. The data were explained in terms of hydrogen bonding between the solutes and the eluent.

Reversed-phase high-performance liquid chromatography (RPLC) has been used to investigate enthalpy–entropy, using octylsilica stationary phase.[26] The compensation temperatures were determined for this system, and the results indicate that their change with the composition of the mobile phase is quite close to that with octadecylsilica stationary phase. It can be concluded that the retention mechanisms of the separation of alkyl benzenes are identical in both systems with the mobile phase exceeding 20% water content.

The separation of substituted benzene derivatives on a reversed-phase C-18 column has been examined.[27] The correlations between the logarithm of the capacity factor and several descriptors for the molecular size and shape and the physical properties of a solute were determined. The results indicated that hydrophobicity is the governing factor controlling the retention of substituted benzenes. Their retention in RPLC is predictable with the guidance of the equations derived by multicombination of the parameters.

It has been found that retention in RPLC is related to the van der Waals volume, pi energy, and hydrogen-bonding energy effects. Nonetheless, higher molecular weight compounds are retained more strongly than anticipated.[28] To investigate this effect more completely, the retention times of phenols were measured on an octadecyl-bonded silica gel in acidic acetonitrile/water mixtures at various temperatures. The enthalpies of phenols were then determined from their log K' values. The extent of the enthalpy effect intensified with increasing molecular size, but the polarity of the molecule was the dominant factor in the enthalpy effect.

Increasing the number of methylene units in alkylbenzene did not extensively influence the pi-energy effect on their retention, but the enthalpy effect did increase significantly.[29] This indicates that a hydrophobic compound can be adsorbed directly onto an octadecyl-bonded silica gel. The value of enthalpy effect of a methylene unit in alkylbenzene was calculated to be 500 cal/mol.

The relationship between solute retention and mobile phase composition in RPLC over the entire range of composition, with emphasis on mobile phases with a high water content, has been studied by Schoenmakers and others.[30] It seems that a quadratic relationship between the logarithm of the capacity factor and the volume fraction of organic modifier is in general valid for mobile phases having less than 90% water. When more water is added to the mobile phase, the quadratic equation becomes insufficient. A study of ten solutes and three organic modifiers is used to show that an extension of the quadratic equation by a term proportional to the square root of the volume fraction leads to a description of the experimental retention data within roughly 10%.

The polarity values of binary acetonitrile/water and methanol/water mobile phases utilized in RPLC were measured and related to methylene selectivity (α_{CH_2}) for both traditional siliceous bonded phases and for a polystyrene–divinylbenzene resin reversed-phase material.[31] The difference in methylene selectivity for both was found to correlate best with percent organic solvent in methanol/water mixtures, while the polarity value gave the best correlation in acetonitrile/water mixtures. The polymeric resin column provided higher methylene selectivity than the siliceous bonded phase at all concentrations of organic solvent.

Measured on the alkylarylketone scale, the retention indices of a set of column test compounds (toluene, nitrobenzene, p-cresol, 2-phenylethanol, and N-methylaniline) were used to determine the changes in selectivity of a series of ternary eluents prepared from methanol/0.02 M phosphate buffer pH 7 (60:40), acetonitrile/0.02 M phosphate buffer pH 7 (50:50) and tetrahydrofuran/0.02 M phosphate buffer pH 7 (25:65). The analyses were carried out on a Spherisorb ODS reversed-phase column. The selectivity changes were often nonlinear between the binary composition.[32]

Direct measurement of solute sorption–desorption kinetics in chromatographic systems gives some beneficial insights into the mechanism of the sorption process and a sensitive means of measuring slight differences in those stationary phase–solvent interactions that are responsible for deciding on the chemical contributions of the stationary phase.

Deviations in retention and selectivity have been studied in cyano, phenyl, and octyl reversed bonded phase HPLC columns. Toluene, phenol, aniline, and nitrobenzene retention in these columns have been measured utilizing binary mixtures of water and methanol, acetonitrile, or tetrahydrofuran mobile phases in an effort to determine the relative contributions of proton donor–proton acceptor and dipole–dipole interactions in the retention process. Retention and selectivity in these columns was compared with polar group selectivities of mobile phase organic modifiers and the polarity of the bonded stationary phases. Notwithstanding the prominent role of bonded-phase volume and residual silanols in the retention process, each column showed some unique selectivities when used with different organic modifiers.[33]

The physicochemical framework has been investigated by comparing the predictions of two models for the combined effects of the composition of the hydroorganic mobile phase and the column temperature on the retention of n-alkylbenzenes on hydrocarbonaceous bonded stationary phases. The "well-mixed" model leads to expressions for the dependence of retention on three factors equivalent to those derived previously from linear extrathermodynamic relationships. The "diachoric" model stems from the assumption that the mobile phase is just the same as the retention model most widely utilized in chromatography with polar sorbents and less polar solvents. Over finite ranges of mobile-phase composition and temperature, each model describes retention behavior. However, only the well-mixed model depicts retention well over the whole range of mobile-phase composition and temperature studied in this situation. The success of the well-mixed model and the limits of the model provide some understanding of the role of organic solvents in determining the extent of chromatographic retention on a nonpolar stationary phase with hydroorganic eluents.[34]

When the intracolumn effect of mass transfer and diffusion is the primary factor affecting band broadening, the column efficiency decreases with the increase of the viscosity of the methanol/water mixture. Conversely, when the extra-column effect is the primary factor, increased viscosity of the eluents will aid in improving column efficiency. Column efficiency is also related to the characteristics of the sample.[35]

C. Solubility and Retention in HPLC

The solubility parameter concept was established in the 1930s in studies by Hildebrand and Scatchard. The original concept includes regular solutions; that is, solutions that show no excess entropy effect on mixing. The solubility parameter theory offers the following interesting features:

1. The basis of the concept is the assumption that the characteristics of mixtures may be described by the properties of pure components. Consequently, the arithmetic expressions involved (regular mixing rule) are relatively simple.
2. The solubility parameter concept relates to compounds rather than to molecules. It is correlated with practical data more conveniently than a molecular statistical approach does because it is a macroscopic approach.

Earlier work on the applicability of the solubility parameter theory to liquid chromatography focused attention on quantitation; for this work, the model has been unsuccessful. Schoenmakers and others[36] believe that the potential of the solubility parameter can help in designing a genuine framework for retention behavior in liquid chromatography. Based on this work, these conclusions have been made:

1. Reasonable retention times are achieved if the polarity of the solute is approximately intermediate between the polarities of the mobile and stationary phases.
2. For higher members of a homologous series with about the same polarity, the logarithm of the capacity factor varies linearly with the molar volume or the carbon number.
3. The absolute difference in polarity between the mobile phase and the stationary phase may be defined as the general selectivity of an HPLC system.
4. There are two frequently used methods to elute a particular compound in HPLC: the normal-phase mode ($\delta s > \delta m$) and the reversed-phase mode ($\delta m > \delta s$). Reversed-phase systems present superior general selectivity. Solutes are eluted in descending order of polarity in reversed-phase systems and in ascending order of polarity in normal-phase systems.
5. Though stationary phases of intermediate polarity (alumina, silica, carbon) afford only moderate general selectivity, they are potentially most powerful for very polar solutes when utilized in the reversed-phase mode.

6. Perfluorinated stationary phases probably offer superior selectivity when compared with the current hydrocarbon-bonded stationary phases.
7. Particular separation effects can be understood from the multicomponent solubility parameter theory. Specific effects for nonpolar compounds are predictable with perfluorinated and graphitized carbon black stationary phases. Specific selectivity for polar compounds in RPLC can be attained with polar additives to the mobile phase.
8. Earlier formulated transfer rules for binary mobile phases in RPLC may be explained by solubility parameter expressions.

VIII. STATIONARY PHASE EFFECTS

The stationary phase effects of adsorption/normal phase or the reversed-phase methods are discussed in the next two sections.

A. Adsorption/Normal Phase

Utilizing molecular liquid adsorption chromatography for the separation of substances is structured on the differences in adsorption from solutions and is determined by the intermolecular interactions of the substance being investigated (solute) and adsorbent, or the eluent (solvent) and the adsorbent, and of the solute and the eluent. The last interaction must be considered both at the interface and in the bulk solution. Depending on the contribution of each of these intermolecular interactions and adsorbents such as silica gel with a hydroxylated surface (commonly used in the normal-phase chromatography), these adsorbents may also be suitable as the adsorbent for the reversed-phase chromatography. Since adsorption from solutions is affected to a great extent by the intermolecular interactions of the solute and the eluent with the adsorbent, surface chemistry attributable to the chemical modifications of adsorbents is important.[37]

The effects of the surface chemistry of the adsorbent, the eluent composition, and the intermolecular interactions on the retention value in liquid chromatography have been studied.[38] The retention volumes of solutes with different molecular structures have been determined on the same silica surface modified by different functional groups from a variety of solvents. The examination of the influence of temperature on the retention volumes permits the assessment of other thermodynamic adsorption characteristics that affect the intermolecular interactions at adsorption from solution.[39,40]

In a liquid–liquid chromatographic system, the interface between the phases can be involved in the distribution process. So, it is necessary to consider the distribution at

- the interface between solid support and the stationary liquid phase
- the stationary liquid phase
- the interface between the stationary liquid phase and the mobile phase
- the mobile liquid phase

It is important to recognize that the following three distribution equilibria may have effects on retention:

- partition between the bulk phases
- adsorption on the solid surface
- adsorption on the liquid–liquid interface

B. Reversed Phase

Reversed-phase (RP) silicas are remarkably useful in answering many separation problems in modern liquid chromatography. Nevertheless, problems still exist with the reproducibility of chromatographic results brought about with these materials. One substantial obstacle to the improvement of RP silicas, with respect to constant properties and quality, could be the lack of dependable analytical methods giving detailed information about the silica matrix itself and about the ligand attached to the surface. The structure of the ligand and its purity, its degree of substitution, and its distribution on the surface of a mono-, di-, and trifunctional bound alkyl groups, and mono- and polymeric substitution are factors of interest.[41] Other important parameters include the proportion of unaltered silanol groups and the extent of end-capping.

The amine interactions with the chromatographic support are complex. The curvature in the k' versus amine concentration plot demonstrates that more than one mechanism is functioning. The behavior of amines is similar to that of alcohols, phenols, and protonated thiols. The latter groups are readily chromatographed by silica-based RPLC. The fact that all these groups are poor hydrogen-bond acceptors but good donors suggests that silanol groups are strong hydrogen-bond donors and poor acceptors, as would be anticipated from their Brønsted acidity. Therefore, depending on the pK_a of the silanol groups and on the mobile phase pH and ionic strength, both hydrogen bonding and ionic interactions may occur with the amine functional groups.

IX. A CASE STUDY OF RETENTION MECHANISM INVESTIGATIONS

A combination of chromatography modes may provide excellent resolution for those components that chromatograph inadequately with a single mode.[42,43] The chromatographic separation of baclofen (I), 4-amino-3-(p-chlorophenyl)-butyric acid, from its potential transformation product (II) with dual-mode chromatography entailing ion-pair reversed-phase chromatography and reversed-phase chromatography, respectively, clearly confirms this point (Table 7).

Ion pairing between the amino group of baclofen and pentane sulfonic acid is essentially accountable for the chromatographic behavior of baclofen on a reversed-phase octadecylsilane column. The transformation product (II), however, does not form an ion pair with pentane sulfonic acid and is, accordingly, separated basically by a reversed-phase partition process. It has

TABLE 7 Effect of Sulfonic Acid on Retention Time of Baclofen

Concentration of C-5 reagent	Retention time (min)	
	Baclofen (I)	Lactam (II)
0.0 M	5.6 (asymmetrical peak)	22.0
0.007 M	10.1 (symmetrical peak)	24.0

Structures of I and II are shown in Table 9.

TABLE 8 Effect of Increasing Concentration of C-5 Sulfonic Acid

Concentration of C-5 reagent (mM)	t_R Baclofen (min)	Calc. Δt_R/mM
0.0	5.97	–
0.22	6.04	0.32
2.2	9.25	1.49
5.4	10.87	0.91
10.8	12.08	0.57
16.2	12.90	0.43
21.6	13.43	0.34
32.3	14.23	0.26
43.1	15.06	0.21

been observed that peak symmetry and analysis time of baclofen (I) can be considerably influenced by the concentration of C-5 sulfonic acid (Table 8). These studies show the significance of investigations on stationary phase dynamics in separations by ion-pair RPLC.

Ionic compounds such as baclofen (I) can be chromatographed by several other modes in HPLC. It is possible to chromatograph it as a cation or an anion on cationic or anionic exchange columns, respectively, or as an ion-pair with the oppositely charged counterion by RPLC. Ionization suppression techniques may also be used in RPLC. The separation mechanisms concerned in the chromatography of baclofen (I) from its transformation product have been investigated by comparison of ion-exchange and ion-pair RPLC. This research was directed primarily at the amino group, which was protonated for the cation exchange chromatography or ion-paired with pentane sulfonated anion for the reversed-phase chromatography. Model compounds were used to judge selectivity for ionic and nonionic compounds (see Table 9 for selectivity of ion-exchange methods). The investigation clearly showed that retention in ion-exchange chromatography can be greatly influenced by the nonionized portion of the molecule; for example, see retention data of compounds I, III, and VI in the table.

TABLE 9 Selectivity of Ion-Exchange Method

Compound	Structure	Retention volume
Lactam		1.97
4-(p-Chlorophenyl) glutarimide		1.97
o-Chlorohydrocinnamic acid		1.97
Baclofen		5.91
Benzylamine		9.84
dl-α-Methylbenzylamine		18.6

X. MOLECULAR PROBES/RETENTION INDEX

An assortment of compounds have been used as molecular probes to evaluate HPLC columns and to characterize them:

- Nonpolar compounds, e.g., benzene, naphthalene
- Polar compounds, e.g., hydroquinone or steroids
- Chelating compounds, e.g., acetyl acetone
- Quaternary compounds, e.g., quaternary ammonium compounds
- Basic compounds, e.g., amines
- Acidic compounds, e.g., toluic acid

Unfortunately, none of the popularly used molecular probes are capable of evaluating column-to-column variation.[44] The absolute prediction of retention of any compound includes the use of a rather complex equation[45,46] that requires the knowledge of various parameters for both the solute and the solvent.[47] The relative prediction of retention is made on the basis of the existence of a calibration line describing the linearity between log k^* and interaction index. This second approach, although less common than the first, is easier to use in practice, and often gives a more accurate figure than the first. Using a suitable choice of calibration solutes, it is feasible to take into account subtle mobile phase effects that cannot be included in the theoretical treatment.

However, certain factors must be authenticated before using a prediction model based on a calibration of the chromatographic system. First, it is essential to limit the number of calibration solutes. It is apparent that using a large number of calibration standards can give a great degree of accuracy, but this is too time-consuming. The use of five to six solutes appears to offer a reliable compromise between accuracy and convenience. Second, one must be able to use these calibration compounds in a somewhat liberal range of solvent compositions. Finally, calibration compounds should be "simple" chemicals, stable, and readily available in any chemical laboratory.

Simple linear equations have been developed for accurately forecasting RPLC retention indices and resolution by the use of steroids as solutes and 2-keto alkanes as reference compounds.[48] These equations have usable applications in predicting whether given pairs of compounds may be separated under specific conditions, or for predicting the conditions that will separate mixtures of compounds in the least amount of time. The procedure may be used to enhance isocratic or gradient separations of compound mixtures. The details of this study demonstrate that the Snyder solvent selectivity triangle concept for characterizing mobile-phase liquids does not consistently group solvents according to selectivity for separating steroids. Contrary to theory, experimental separations frequently differ considerably within a particular solvent group. Selectivity differences between solvents in the same group occasionally exceed those between solvents in different groups.

A study has been carried out to utilize the 2-keto alkane system for calculating HPLC retention indices of a series of steroids, and then to use the retention index data to predict retention and resolution of the compounds as a function of solvent strength and selectivity. The primary goal was to utilize this ability to predict resolution for optimizing the separation of steroid mixtures or pairs of individual steroids. A secondary purpose was to investigate the selectivity characteristics of solvents grouped according to the solvent selectivity triangle concept for their capacity to separate particular pairs of steroids.

These results challenge the theory of the solvent selectivity triangle concept (Section XI), which states that solvents in the same group should result in similar selectivity, while those in different groups should render different selectivities.[49,50] A literature search showed widespread usage of the selectivity triangle as a rationale for solvent selection. Nonetheless, with the exception of a publication by Lewis and others,[51] definitive studies on the accuracy of the solvent groupings in the selectivity triangle are lacking. These authors investigated the separation of polystyrene oligomers, using a total of 17 solvents representing all eight of the selectivity groups, and they concluded that the solvent triangle did not accurately predict selectivity for the separation being studied. According to these researchers, the degree of the solute solubility in the pure mobile phase solvents was a better predictor of selectivity than were the groupings of the solvent triangle.

Studies with steroids and polystyrene oligomers demonstrated that solvents classified within the same Snyder selectivity group do not inevitably result in similar selectivity. This inconsistency might be because of the underlying assumptions of the solvent triangle theory, which takes for granted that selectivity is governed to a great extent by the ability of the solvent to engage in hydrogen bonding and dipolar interactions and that dispersion interactions do not have an important role in selectivity for solutions of polar solvents.

The solvent triangle concept also does not consider the role of the stationary phase and the nature of the solutes themselves in influencing a particular separation. Additionally, this concept assumes that only three test solutes are necessary in order to determine the primary selectivity characteristics of various solvents.

A method has been presented to characterize variations in the retention properties of RPLC by column-eluent combinations by utilizing retention indices of a set of reference compounds, toluene, nitrobenzene, p-cresol, and 2-naphthylethanol.[52] These compounds were chosen by multivariate analysis to yield the maximum discrimination between eluents and columns.

An interesting pair of compounds is caffeine and theophylline[53]; these compounds are relatively polar compounds with different functional groups (tertiary and secondary amine). In a few cases, more appropriate comparisons have been made, such as those between androstenedione/testosterone and methyl benzoate/anisole; these compounds are expected to be in different Snyder interaction groups.

Some researchers have used a proposed ASTM test mixture consisting of benzaldehyde, acetophenone, methyl benzoate, dimethyl terphthalate, benzyl alcohol, and benzene to illustrate separation on a column.[54] The first four compounds, however, are from the same interaction group and should act in the same way on changing conditions. The first three have almost constant indices (760, 800, and 890, respectively) so that they, in effect, produce an "index scale" with constant differences against which the last two compounds may be compared.[52,55]

To determine the suitability of retention indices, based on the alkylarylketone scale as the foundation of a reproducible procedure for reporting retentions, the separation of ten barbiturates and a set of column test compounds were investigated on an octadecylsilyl bonded silica (ODS-Hypersil) column with methanol buffer of pH 8.5 as eluent.[56] The effects on the capacity factors and retention indices, on changing the eluent composition, pH, ionic strengthened temperature, indicated that the retention indices of the barbiturates were much less susceptible to small alterations in the eluent than the capacity factors. The retention indices were essentially independent of the experimental conditions in the case of nonionized compounds.

The silanophilic quality of 16 RPLC columns has been evaluated with dimethyldiphenylcyclam, a cyclic tetraaza macrocycle.[57] The method is rapid, does not necessitate the removal of packing material, and uses a water-miscible solvent. The results demonstrate two points: first, cyclic tetraaza macrocycles present substantial benefits over currently used silanophilic agents. Second, the method can readily distinguish the performance of various columns in terms of their relative hydrophobic and silanophilic contributions to absolute retention.

A mixture of 1-nitronaphthalene, acetylacetone, and naphthalene has been suggested for evaluating reversed-phase packing material.[58] This exhibits the usual optimum kinetic chromatographic parameters (the naphthalene peak), the degree of activity or end-capping status of the column (the ratio of the 1-nitronaphthalene and naphthalene retention times), and trace metal activity (the shape and intensity of the acetylacetone peak).

XI. MOBILE PHASE SELECTION AND OPTIMIZATION

The conventional approaches to mobile phase selection and optimization are discussed here. The primary focus is on separations involving compounds with molecular weight less than 2000. More detailed information including coverage of macromolecules may be found in some basic texts.[1,3,59,60] Various modes of chromatography utilized to separate these compounds may be classified as follows:

1. adsorption chromatography
2. normal-phase chromatography
3. reversed-phase chromatography
4. ion-pair chromatography
5. ion-exchange chromatography
6. ion chromatography

Ion-pair chromatography is frequently performed in the reversed-phase mode and is therefore discussed in Section XI. B. Since ion chromato- graphy is an offshoot of ion-exchange chromatography, it has been discussed right after ion-exchange chromatography (Section XI. B).

A. General Considerations

A variety of means have been used to optimize separations for each chromatographic approach (see Chapters 5–8 in Reference 3); the discussion here is limited to conventional approaches used to select the mobile phase.

These approaches are often based on intuitive judgment and the know-how of the chromatographer. For the latter, it is important to underscore the fact that knowledge of the physicochemical basis of retention and basic understanding of separation mechanisms in HPLC, discussed earlier, will help tremendously in choosing the correct mobile phase quickly, then optimizing it by the usual experimentation. These tests can be logically conducted by having a computer select mobile-phase combinations based on certain preset requirements. One can appreciate better this aspect of method development when the conventional approaches have been mastered.

1. Sample Properties

The choice of an HPLC method should be made principally from the properties of the sample (alternate terms are solute, analyte, or eluite) once it has been determined that it has a sufficiently low molecular weight (i.e., less than 2000) to warrant use of the techniques mentioned earlier. The decision could be made on the basis of the solubility of the sample, that is, whether it is soluble in polar or nonpolar solvents (Table 10).

From Table 10, it is apparent that compounds having a molecular weight greater than 2000 are better separated with gel-permeation chromatography or size-exclusion chromatography (SEC). SEC can, of course, be used for molecular weights below 2000 just as other modes of chromatography may be used for compounds with higher molecular weight. The discussion here is concerned essentially with compounds of molecular weight greater than 2000.

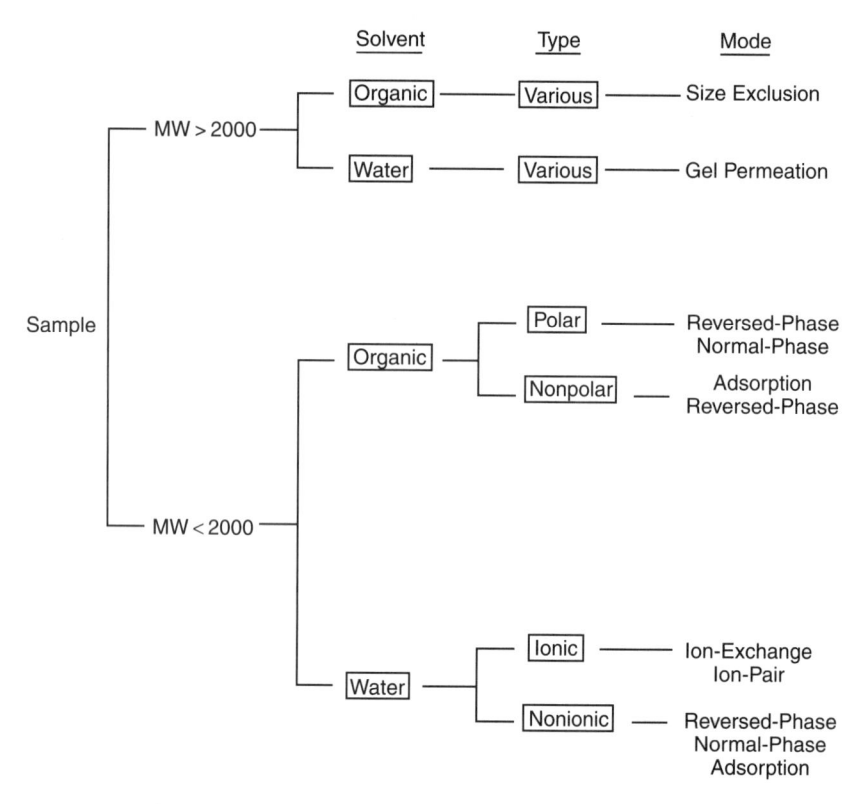

TABLE 10 Selection of Chromatographic Method Based on Solubility

```
                          Solvent         Type            Mode

                       ┌─ Organic ──── Various ──── Size Exclusion
              ┌ MW>2000┤
              │        └─ Water ────── Various ──── Gel Permeation
              │
              │                        ┌─ Polar ──── Reversed-Phase
              │                        │            Normal-Phase
              │        ┌─ Organic ─────┤
              │        │               └─ Nonpolar ─ Adsorption
   Sample ────┤        │                            Reversed-Phase
              │        │
              └ MW<2000┤
                       │               ┌─ Ionic ──── Ion-Exchange
                       │               │            Ion-Pair
                       └─ Water ───────┤
                                       └─ Nonionic ─ Reversed-Phase
                                                    Normal-Phase
                                                    Adsorption
```

Knowing whether the sample is ionic can help in determining the suitable mode of chromatography. In this regard, the dissociation constant of the compound is of great value—because with appropriate adjustment of pH, one can select a desirable percent ionization of the compound of interest, remembering that when $pH = pK_a$, the compound is 50% ionized.

2. Column Selection

The choice of column should be made after thorough consideration of mode of chromatography, column-to-column variability, and a number of other considerations.[3-5] A brief discussion on columns and column packings is given here. The column packings may be classified according to the following features:[1]

- rigid solids or hard gels
- porous or pellicular and superficially porous particles
- spherical or irregular particles
- particle size (dp)

Rigid solids based on a silica matrix are customarily used as HPLC packings. These packings can tolerate the relatively high pressures (10,000–15,000 psi) associated with the technique. The silica particles may be acquired in a variety of sizes, shapes, and degrees of porosity. In addition, various functional groups or polymeric layers can easily be attached to the silica surface, extending the usefulness of these particles for applications to any specific HPLC method.

Hard gels are commonly based on porous particles of polystyrene cross-linked with divinylbenzene. Depending on the method of preparation, the resulting particles may vary in rigidity and porosity over reasonably wide limits. They still find use in ion exchange and SEC; however, rigid solids are gradually replacing hard gels.

Packings for HPLC may be further characterized as either pellicular or porous. Pellicular particles are fabricated from spherical glass beads and then coated with a thin layer of stationary phase. For instance, a porous layer can be deposited onto the glass bead to generate a porous layer or a superficially porous particle. Then the porous layer can be coated with liquid stationary phase or reacted to give a bonded stationary phase. Pellicular particles are in general less efficient than the porous layer of superficially porous particles.

As the particle size (dp) decreases, column plate height declines and the column becomes less permeable. Consequently, for small values of dp, a column of some fixed length produces higher plate count; that is, higher efficiency, but greater pressure drop across the column is necessary for a particular value of linear flow. This indicates that columns of small particles are more efficient, but demand higher operating pressures.

For columns of similar efficiency, the maximum sample size allowed is generally smaller when using pellicular particles than when using porous particles. The reason for this is that less stationary phase is available per unit volume of column. Approximately five times as much sample can be charged to a porous column before there is a notable decrease in k'. Since bigger samples can be injected onto a porous-particle column, the resulting bands are larger and more readily detected. The columns of small porous particles allow good detection sensitivities and are favored for ultratrace analysis (3a).

To compare similar columns, it is important that experimental conditions for the test chromatogram are painstakingly reproduced and that sufficient time is allowed for column equilibration before starting the test. The anticipated changes in column performance parameters related to changes in the experimental conditions are summarized in Table 11. Table 12 lists the columns commonly used in HPLC, along with the mode of separation, particle size, and functionalities.

3. Column Evaluations

As mentioned earlier, columns must be thoroughly evaluated prior to use.[3] Some of the desirable properties of test solutes are listed here.

1. Test solutes should be of low molecular weight to guarantee rapid diffusion and effortless access to the packing pore structure.
2. Test solutes should have strong absorbance, preferably at 254 nm.
3. The solute should include components that

    (a) characterize the column in terms of both kinetic and thermodynamic performance

TABLE 11 HPLC Parameters Affecting Column Efficiency

Parameter	Change in efficiency (N)
Flow rate	Low flow rate generally gives high value of N
Particle size	Small particle size gives high value of N
Column length, L	N is proportional to L
Mobile phase viscosity	Low value gives high value of N
Temperature, T	High values reduce viscosity and give high values of N
Capacity factor, k'	Low k' (<2) give low values of N; for high k' values (>2), N is influenced
Dead volume	N is decreased because of band-broadening contributions to peak width
Sample size	Large amounts (mg) or large volumes decrease N

TABLE 12 HPLC Columns

Mode	Material	Particle size (μM)	Treatment
Adsorption	Silica, irregular	2–20	Unreacted
	Silica, spherical	5–10	Unreacted
	Alumina, irregular	3–12	Unreacted
	Alumina	5–20	Unreacted
Reversed-phase	Silica with long C chain	3–15	C-18
	Silica with intermediate C chain	5–10	C-8
	Silica with short C chain	5–10	C-1, C-3
Normal-phase	Silica (weak)	5–15	Ester, ether, diester
	Silica (medium)	5–15	NO_2, CN
	Silica (high)	5–15	Alkylamino, amino
Ion-exchange	Silica (anion)	5–15	NMe^{3+}, NR^{3+}, NH_2
	Resin (anion)	7–20	NMe^{3+}, $-NH^{3+}$
	Silica (cation)	5–10	$-SO^{3-}(H^+)$, $-SO^{3-}$, $(NH_4)^+$
	Resin	5–20	$-SO^{3-}$

 (b) determine the column dead volume, i.e., of the test mixture as an unretained solute
 (c) differentiate retention with k' values between 2 and 10

4. Mobile-Phase Selection in HPLC

As mentioned before, the retention in HPLC depends on the strength of the solute's interaction with both the mobile and stationary phases, as opposed to gas chromatography where the mobile phase does not contribute to the selectivity. Therefore, it is important that an intelligent choice of the type of stationary phase for the separation is made, and selectivity is altered by changing the mobile phase. The selection of the mobile phase for a given separation is thus a very important decision in HPLC.

Some rather broad generalizations can be made about the selection of certain preferred solvents from the plentiful ones available for HPLC. An appropriate solvent will have low viscosity, be compatible with the detection system, be easily available in pure form, and, if possible, have low flammability and toxicity. When choosing organic solvents for use in mobile phases, several physical and chemical properties of the solvent should be carefully thought about. The refractive index or UV cutoff values are also important considerations from the standpoint of detection.

The term *polarity* concerns the ability of a sample or solvent molecule to interact by combination of dispersion, dipole, hydrogen bonding, and dielectric interactions (see Chapter 2 in Reference 3). The combination of these four intermolecular attractive forces constitutes the solvent polarity—a measure of the strength of the solvent. Solvent strength increases with polarity in normal-phase and adsorption HPLC and decreases with polarity in reversed-phase HPLC. Consequently, polar solvents preferentially attract and dissolve polar solute molecules.

Common HPLC solvents with satisfactory purity are commercially available. Halogenated solvents can have traces of acidic impurities that react with stainless steel components of the HPLC system. Mixtures of halogenated solvents with water should not be stored for long periods, as they are inclined to decompose. Mixtures of halogenated solvents with various ethers (e.g., diethyl ether) react to form products that are particularly corrosive to stainless steel. Halogenated solvents such as methylene chloride react with other organic solvents such as acetonitrile and, on standing, form crystalline products.

5. Solvent Selection

Snyder has described a scheme for classifying common solvents according to their polarity or chromatographic strength (P' values) based on their selectivity or relative ability to engage in hydrogen bonding or dipole interactions (for more details, see Chapter 2 in Reference 3). Common solvents have been classified into eight groups (I–VIII) showing significantly different selectivities (Fig. 9).

Group	Common Solvents
I	Aliphatic ethers, tetramethylguanidine
II	Aliphatic alcohols
III	Tetrahydrofuran, pyridine derivatives, glycol ethers, sulfoxides
IV	Acetic acid, formamide, benzyl alcohol, glycols
V	Methylene chloride, ethylene chloride
VIa	Aliphatic ketones and esters, dioxane, tricresyl phosphate
VIb	Sulfones, nitriles
VII	Aromatic hydrocarbons, halo-substituted aromatic hydrocarbons, nitro compounds, aromatic ethers
VIII	Water, *m*-cresol, fluoroalcohols, chloroform

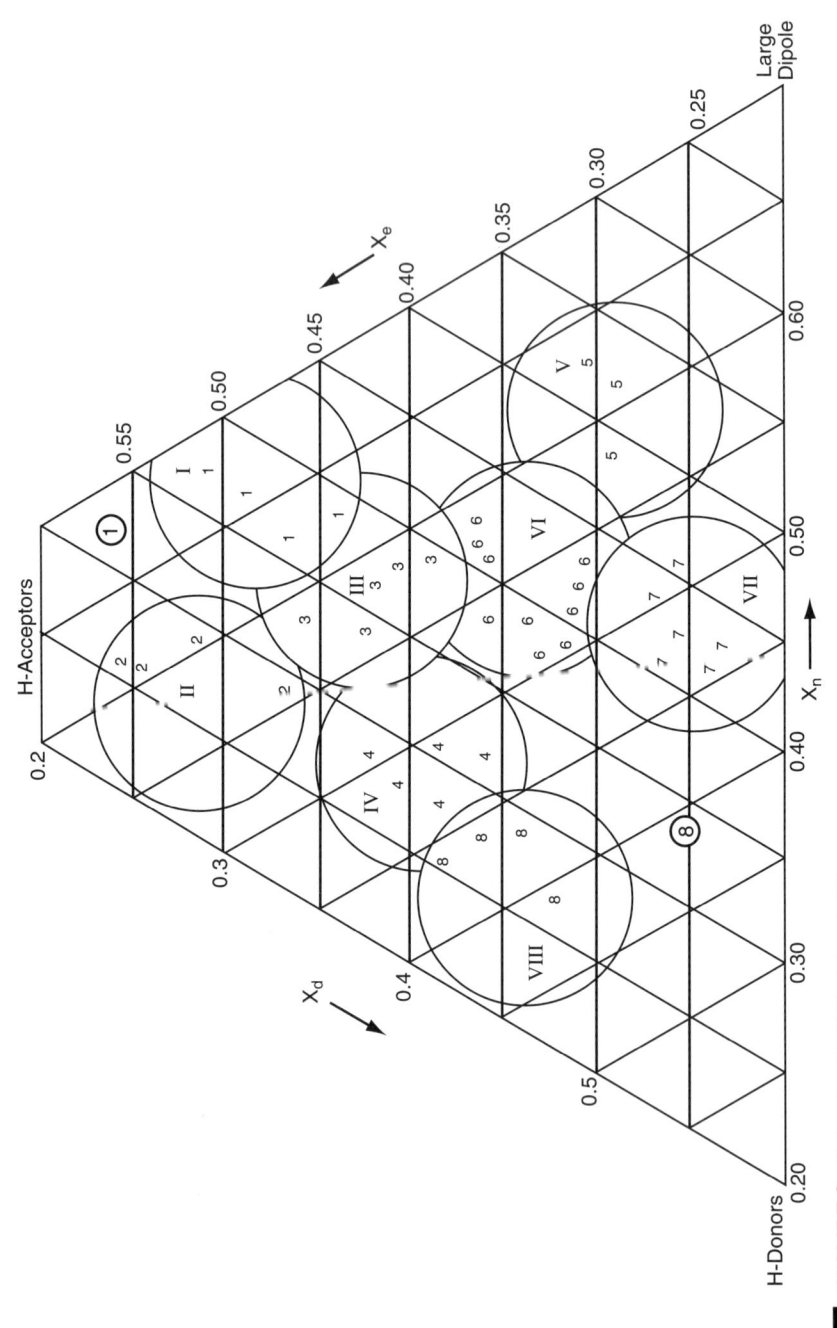

FIGURE 9 Solvent-selectivity triangle (3).

TABLE 13 Polarity of Some Common Solvents

Solvent	δ	P'	ε
n-Hexane	7.3	0.1	0.00
Ethyl ether	7.4	2.8	0.43
Triethylamine	7.5	1.9	
Cyclohexane	8.2	0.2	0.03
Carbon tetrachloride	8.6	1.6	0.11
Ethyl acetate	8.9	4.4	0.48
Tetrahydrofuran	9.1	4.0	0.53
Chloroform	9.3	4.1	0.26
Methylene chloride	9.6	3.1	0.30
Acetone	9.7	5.1	0.53
Dioxane	10.1	4.8	0.51
Dimethyl formamide	11.8	6.4	
Isopropanol	12.0	3.9	0.60
Acetonitrile	12.1	5.8	0.52
Ethanol	12.7	4.3	
Methanol	14.5	5.1	0.70
Formamide	19.2	9.6	
Water	23.4	10.2	

Note: Adapted from Reference 3.
δ = Hildebrand solubility parameter
P' = Polarity index
ε = Solvent strength for silica adsorbent

The P' (polarity index) values and selectivity group classifications for some solvents commonly used in HPLC are given in Table 13.

The values of P' and selectivity factors are determined from the experimentally derived solute polarity distribution coefficient for the test solutes ethanol, dioxane, and nitromethane. The solute distribution coefficients are corrected for effects related to solute molecular size, solute/solvent dispersion interactions, and solute/solvent induction due to solvent polarizability. The resulting parameters P' and solvent selectivity should reflect only the selective interaction properties of the solvent. The test solutes ethanol, dioxane, and nitromethane are utilized to gauge the strengths of solvent proton acceptor, proton donor, and strong dipole interactions, respectively.

Modifications in the mobile phase can result in important selectivity changes for various sample analytes. The greatest change in mobile-phase selectivity can be obtained when the relative signficance of the various intermolecular interactions between solvent and solute molecules is considerably altered. The changes in selectivity may be affected by making use of the following solvent properties:

- Proton acceptors: amines, ethers, sulfoxides, amides, esters, and alcohols
- Proton donors: alcohols, carboxylic acids, phenols, and chloroform

- Large dipole solvents: methylene chloride, nitrites, sulfoxides, and ketones

It is hard to anticipate that replacement of methanol by another alcohol such as propanol would radically change selectivity, because in both cases a proton-donor solvent is present. However, a greater change in selectivity would be expected by using ethyl ether (proton acceptor) or methylene chloride (large dipole moment).

The solvent classification scheme is advantageous when identifying solvents with different chromatographic selectivities. It is generally preferable to use mixtures of solvents rather than just a single pure solvent as the mobile phase. For binary solvents, combining a strength-adjusting solvent with various volume fractions of a strong solvent makes possible the complete polarity or solvent strength range between the extremes represented by the pure solvents themselves to be covered. The strength-adjusting solvent is generally a nonselective solvent, such as water, for reversed-phase chromatography and hexane for normal-phase techniques. The solvent strength of a binary solvent combination is the arithmetic average of the solvent strength weighting factors modified according to the volume fraction of each solvent. For normal-phase chromatography, the solvent strength-weighting factor, S_i, is the same as the polarity index, P'.

In reversed-phase chromatography, a different set of experimental weighting factors is used.[61]

The solvent strength for any solvent mixture can be calculated from the following equation:

$$S_T = \sum_i S_i \theta_i \qquad (22)$$

where S_T = total solvent strength of the mixture, S_i = solvent strength weighting factor, and θ_i = volume fraction of solvent in the mixture.

Binary solvent mixtures give a simple means of regulating solvent strength but limited opportunities for controlling solvent selectivity. It is possible, when using ternary and quaternary solvent mixtures, to fine-tune solvent selectivity while maintaining a constant solvent strength.[62–64] In addition, there are only a small number of organic modifiers that can be used as binary mixtures with water.

The Snyder solvent selectivity triangle concept may be combined with a mixture-design statistical technique to define the optimum mobile phase composition for a given separation. A characteristic of this mixture-design technique is that it leads to the choice of a quaternary mobile-phase system for most separations. The selection process can be controlled by a microprocessor in an interactive manner if the solvent delivery system can pump four solvents concurrently (for more details, see Chapter 11 in Reference 3).

6. Mobile-Phase Additives

It is necessary at times to add reagents such as buffers, ion-pairing reagents, or other modifiers such as triethylamine to the mobile phase to improve reproducibility, selectivity, or peak shape. Buffers are used primarily to regulate the pH and the acid–base equilibrium of the solute in the mobile phase. They may also be utilized to affect the retention times of ionizable compounds. The buffer capacity should be at the maximum and should not vary in the pH range of 2 to 8 commonly used in HPLC. The buffers should be soluble, stable, and compatible with the detector employed; for example, citrates are known to react with certain HPLC hardware components and their use should be avoided in these situations.

Adding compounds such as long-chain alkyl compounds in reversed-phase separations will affect the retention of ionic compounds but will have no effect on nonionic compounds unless the concentration is sufficiently high to form micelles. For more information, please refer to Chapter 8 in Reference 3.

In reversed-phase separations of basic compounds, competing amines such as triethylamine and di-*n*-butylamine have been added to the mobile phase to improve peak shape. Acetic acid can serve a similar purpose for acidic compounds. These modifiers, by competing with the analyte for residual active sites, cause reduction of retention time and peak tailing.

Other examples of mobile phase modifications are the addition of silver ions to separate geometric isomers and the inclusion of metal ions with chelating agents to separate racemic mixtures.

B. Modes of Chromatography

Details on mobile-phase selection for each of the modes of chromatography are discussed below; included are basic information on mechanism, solvent and column selection, and other practical considerations. Reversed-phase technique is discussed at greater length since it is the most commonly used technique in HPLC.

1. Adsorption Chromatography

Adsorption chromatography using polar stationary phases with relatively nonpolar mobile phases is sometimes referred to as "liquid–solid" chromatography. Separations in adsorption chromatography result to a great extent from the interaction of sample polar functional groups with discrete adsorption sites on the stationary phase. The strength of these polar interactions (see Table 14) is accountable for the selectivity of the separation in liquid–solid chromatography.

Therefore, liquid–solid chromatography is usually considered appropriate for the separation of nonionic molecules that are soluble in organic solvents. Highly polar compounds, those with high solubility in water and low solubility in organic solvents, interact quite strongly with the adsorbent surface, and as a result, one obtains peaks of poor symmetry and poor efficiency.

Solvent strength and selectivity can be conveniently managed by using binary or higher-order solvent mixtures. The alteration in solvent strength as a function of the volume percent of the more polar component is not a linear function. At low concentrations of the polar solvent, modest increases in concentration produce large increases in solvent strength. At the other extreme, comparatively large changes in the concentration of the polar solvent have an

TABLE 14 Classification of Functional Group Adsorption Strengths

Retention	Type of compound
Nonadsorbed	Aliphatics
Weakly adsorbed	Alkenes, mercaptans, sulfides, aromatics, and halogenated aromatics
Moderately adsorbed	Polynuclear aromatics, ethers, nitrites, nitro compounds, and most carbonyls
Strongly adsorbed	Alcohols, phenols, amines, amides, imides, sulfoxides, and acids
Comparative	F < Cl < Br < I; cis compounds are retained more strongly than trans; equatorial groups in cyclohexane derivatives (and steroids) are more strongly retained than axial derivatives

Note: Adapted from Reference 59.

effect on the solvent strength of the mobile phase to a lesser degree. When the optimal solvent strength has been determined for a separation, changing solvent selectivity at a constant solvent strength enhances the resolution of the sample.

The uptake of water by the column adversely affects separation and leads to irreproducible separations and long column regeneration times. Using a 100% saturated solvent is not desirable because such liquid–solid chromatographic systems are frequently unstable. Evidently, in this situation the pores of the adsorbent gradually fill with water, leading to shifts in retention with time and perhaps also to a change in the retention mechanism as liquid–liquid partition effects become more significant. Fifty percent saturation of the mobile phase has been advocated for stable chromatographic conditions when silica is the adsorbent.[65–67] Solvents with 50% water saturation can be made up by combining dry solvent with a 100% saturated solvent or, preferably, by using a moisture-control system.[68] The latter is composed of a water-coated thermostatically controlled adsorbent column through which the mobile phase is recycled until the desired degree of saturation is attained.

A column that has been deactivated with water might not display adequate separation properties. Restoring the activity of the column by forcing a large amount of dry mobile phase through the column is a tedious and expensive process. As an alternative, reactivation may be achieved chemically by using the acid-catalyzed reaction between water and 2,2-dimethoxypropane, the products of which (acetone and methanol) are easily eluted from the column.[69]

Besides water, virtually any organic polar modifier can be used to control solute retention in liquid–solid chromatography. Alcohols, acetonitrile, tetrahydrofuran, and ethyl acetate in volumes of less than 1% may be merged into nonpolar mobile phases to control adsorbent activity. Overall, column efficiency decreases for alcohol-modulated eluents compared to water-modulated eluent systems.

The retention behavior of a sample solute is determined to a great extent by its functional groups. Consequently, adsorption chromatography has proven useful for different classes of compounds. Isomeric and multifunctional groups can usually be separated by adsorption chromatography because the relative position of the solute functional groups interacting with the spatial arrangement of the surface hydroxyl groups controls adsorption. This characteristic leads to a prominent selectivity of silica gel for positional isomers.

Organic amine compounds can be separated on silica gel columns with good peak symmetry when using organic/aqueous mobile phases.[70–72] Solute retention appears to involve both electrostatic and adsorption forces.

a. Normal bonded phases

Polar bonded phases that have a diol, cyano, diethylamino, amino, or diamino functional group are commercially available; representative structures as well as applications are shown in Table 15. The diol- and diethylamino-substituted stationary phases are also useful in size-exclusion and ion-exchange chromatography, respectively.

When used with a mobile phase of low polarity, the alkylnitrile- and alkylamine-substituted stationary phases behave in a manner similar to the solid adsorbents discussed in the previous section. That is, the retention of the sample increases with solute polarity, and increasing the polarity of the mobile phase diminishes the retention of all solutes. The polar bonded phase packings are in general not as retentive as adsorbent packings, but they are relatively free of the problems of chemisorption, tailing, and catalytic activity associated with silica and alumina. The bonded-phase packings react quickly to changes in mobile-phase composition and may be utilized in gradient elution analyses. Adsorbent packings respond gradually to changes in mobile-phase composition because of slow changes in surface hydration, making gradient-elution analysis difficult. The polar bonded phase packings have been proposed as alternate choices to microporous adsorbents for separating the same sample type because of these advantages.[73]

The alkylnitrile-substituted phase has intermediate polarity and is less retentive than silica gel, but it exhibits similar selectivity. It affords good selectivity for the separation of double-bond isomers and ring compounds

TABLE 15 Polar Bonded Phases

Designation	Structure	Application
Diol	$-(CH_2)_3OCH_2CH(OH)CH_2(OH)$	Surface-modifying groups for silica packings used in SEC
Cyano	$-(CH_2)_3CN$	Partition or adsorption chromatography
Amino	$-(CH_2)_nNH_2$ ($n = 3$ or 5)	Adsorption, partition, or ion-exchange chromatography
Dimethyl-amino	$-(CH_2)_3N(CH_3)_2$	Ion-exchange chromatography
Diamino	$-(CH_2)_2NH(CH_2)_2\ NH_2$	Adsorption or ion-exchange chromatography

Note: Adapted from Reference 59.

differing in either the position or number of double bonds.[74] The alkylnitrile-substituted stationary phases have been used for the separation of saccharides that are poorly retained on reversed-phase columns using aqueous mobile phases. The alkylamine-substituted phases provide a separation mechanism complementary to either silica gel or alkylnitrile-substituted phases. The amino function confers strong hydrogen-bonding qualities, as well as acid or base properties, to the stationary phase, depending on the characteristics of the solute. Class separation of polycyclic aromatic hydrocarbons has been achieved using aminoalkyl-substituted stationary phase.[75,76] Retention is based principally on charge-transfer interactions between the aromatic π electrons of the polycyclic aromatic hydrocarbons and the polar amino groups of the stationary phase. Samples are separated into peaks containing those components with the same number of rings. Retention increases incrementally with increasing ring number, but is barely affected by the presence of alkyl ring substituents. Reversed-phase separations, nevertheless, show poor separation between alkyl-substituted polycyclic aromatic hydrocarbons and polycyclic aromatic hydrocarbons of a higher ring number.

The practice of normal-phase chromatography resembles that described for adsorption chromatography. A polar solvent modifier, such as isopropanol at the 0.5–1.0% v/v level, is used in nonpolar solvents to improve peak symmetry and retention time reproducibility. It is hypothesized that the polar modifier solvates the polar groups of the stationary phase, providing an improvement in mass-transfer properties. Either glacial acetic acid or phosphoric acid is used at low levels as a tailing inhibitor in the separation of carboxylic acids or phenols. Similarly, propylamine is an appropriate modifier for the separation of bases.

Certain problems come about with the use of alkylamine-substituted stationary phases. Because amines are readily oxidized, degassing the mobile phase and avoiding solvents that may contain peroxides (for example, diethyl ether and tetrahydrofuran) are recommended. Samples or impurities in the mobile phase that contain ketone or aldehyde groups may react chemically with the amine group of the stationary phase, forming a Schiff's base complex.[74] This reaction will change the separation properties of the column. Flushing with a large volume of acidified water can regenerate the column.[77]

2. Reversed-Phase Chromatography

Reversed-phase chromatography is carried out on columns where the stationary phase surface is less polar than the mobile phase. The most frequently used column packings for reversed-phase separations possess a ligand such as octadecyl (C-18), octyl (C-8), phenyl, or cyano-propyl chemically bonded to microporous silica particles. The silica particles may be spherical or irregularly shaped and contain unreacted silanol groups. These silanol groups may be end-capped by silanization with a small silanizing reagent such as trimethylchlorosilane (TMCS).

It is estimated that over 65% (possibly as high as 90%) of all HPLC separations are executed in the reversed-phase mode. The rationale for this includes the simplicity, versatility, and scope of the reversed-phase method.[78] The hydrocarbon-like stationary phases equilibrate quickly with modifications in mobile phase composition and are, as a result, well suited for use with gradient elution.

Retention in reversed-phase chromatography occurs by nonspecific hydrophobic interactions of the solute with the stationary phase. The ubiquitous application of reversed-phase chromatography arises from the fact that practically all organic molecules have hydrophobic regions in their structures and effectively interact with the stationary phase. Therefore, reversed-phase chromatography is ideally suited to separating the components of oligomers or homologues. The logarithm of the capacity factor is generally a linear function of the carbon number within a homologous series. Branched chain compounds are ordinarily retained to a lesser extent than their straight-chain analogues and unsaturated analogues. Since the mobile phase in reversed-phase chromatography is polar and usually contains water, the method is perfectly suitable for the separation of polar molecules that are either insoluble in organic solvents or bind too strongly to solid adsorbents for normal elution. Many samples that have a biological origin fall in this category.

It is believed that retention in reversed-phase chromatography is a function of sample hydrophobicity, whereas the selectivity of the separation is almost wholly a result of specific interactions of the solute with the mobile phase.[79] In general, the selectivity can be conveniently modified by changing the kind of organic modifier in the mobile phase. Appropriate pH is used to suppress ionization, or ion-pairing reagents are used to increase lipophilicity to assist the degree of solute transfer to the stationary phase and thus control selectivity for ionic or ionizable solutes.[80] To separate optically active isomers, metal–ligand complexes and chiral reagents may be added to the mobile phase.

The fine points of the mechanism controlling retention in reversed-phase chromatography using chemically bonded hydrocarbonaceous phases are not completely understood.[81] Solute retention in RPLC can progress either by partitioning between the hydrocarbonaceous surface layer of the nonpolar stationary phase and the mobile phase or by adsorption of the solute to the nonpolar portion of the stationary phase. In these circumstances, the partitioning mechanism seems less probable since the hydrocarbonaceous layer is only a monolayer thick and does not have the favorable attributes of a bulk liquid for solubilizing solutes. However, the less polar solvent components of the mobile phase could amass near the apolar surface of the stationary phase, forming an essentially stagnant layer of mobile phase abounding in the less polar solvent.[82] Therefore, solute could partition between this layer and the bulk mobile phase and not directly interact with the stationary phase proper. The remainder of the evidence favors the adsorption mechanism either with the stationary phase surface itself or by interaction with the ordered solvent molecule layer at the stationary phase surface.[83]

To provide a simple view of solvophobic theory, we will assume that solute retention occurs by adsorption of the solute to the stationary phase, without defining the stationary phase. The solvophobic theory assumes that aqueous mobile phases are highly structured because of the tendency of water molecules to self-associate by hydrogen bonding and this structuring is perturbed by the presence of nonpolar solute molecules. As a consequence of this very high cohesive energy of the solvent, the less

TABLE 16 Solvent Strength in Reversed-Phase LC (1)

Solvent	P'	k'*
Water	10.2	
Dimethyl sulfoxide	7.2	1.5-fold
Ethylene glycol	6.9	1.5
Acetonitrile	5.8	2.0
Methanol	5.1	2.0
Acetone	5.1	2.2
Ethanol	4.3	2.3
Tetrahydrofuran	4.0	2.8
i-Propanol	3.9	3.0

*Decrease in k' for each 10% addition of solvent to water

polar solutes are literally forced out of the mobile phases and are bound to the hydrocarbon portion of the stationary phase. Therefore, the driving force for solute retention is not the favorable affinity of the solute for the stationary phase, but is due to the solvent's forcing the solute to the hydrocarbonaceous layer.

The most frequently used solvents for RPLC are methanol, acetonitrile, and tetrahydrofuran used in binary, ternary, or quaternary combinations with water. The effect of solvent strengths can be seen in Table 16. A significant difference in separation selectivity can be achieved by replacing the organic solvent with a different solvent.

Changes in pH can alter the separation selectivity for ionized or ionizable solutes, since charged molecules are distributed preferentially into the aqueous or more polar phase (For further discussion, see Chapter 6, Reference 3). In general, separation conditions that are used to vary α values of various peaks are summarized in Table 17.

TABLE 17 Separation Conditions Used to Improve α Values

Variable	Impact
Stationary phase	Choice of C-18, C-8, phenyl, cyano, or trimethyl has varying impact, depending on type of sample
Organic solvent	Change from methanol to acetonitrile or THF commonly produces large change in α values
pH	Change in pH can have a major effect on α values of acidic or basic compounds
Solvent strength	Changes in percent of organic solvent often provides significant changes in values
Additives	Ion-pair reagents have great impact on α values. Other additives such as amine modifiers, buffers, and salts including complexing agents have various effects
Temperature	Vary between 0 °C to 70 °C to control α values

Simple samples

Initial runs with ACN/water → Define k'-range vs. %-ACN → Select best %-ACN for k'-range and band spacing

Normal samples

Initial runs with MeOH/water → Define k'-range vs. %-MeOH → Select best %-THF for k'-range and band spacing

Initial runs with THF/water → Define k'-range vs. %-THF → Select best %-MeOH for k'-range and band spacing

Difficult samples

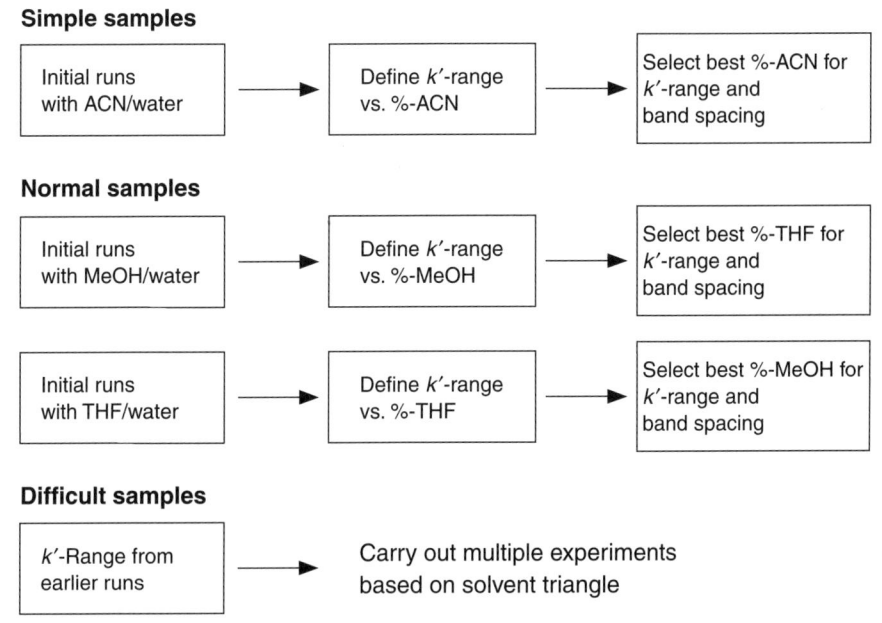

k'-Range from earlier runs → Carry out multiple experiments based on solvent triangle

FIGURE 10 A recommended approach for developing HPLC methods.

Frequently, the choice of the mobile phase equates to methanol/water or acetonitrile mixtures in various proportions. After selecting the mobile phase, the next step is to optimize the concentration of the organic solvent. Following that, low concentrations of tetrahydrofuran are explored to further improve separations (see Fig. 10).

Resolution can be mapped as a function of various proportions of acetonitrile, methanol, and THF in the mobile phase. Usually k' range or run time is held constant during the process by varying the amount of water in the mobile phase mixture to compensate for small differences in the strength of the three pure organic solvents. If further improvement in separations is needed, the additives given in Table 18 should be considered.

3. Ion-Pair Chromatography

Ionic or partially ionic compounds can be chromatographed on reversed-phase columns by using ion-pairing reagents. These reagents are typically long-chain alkyl anions or cations that, when used in dilute concentrations, can increase the retention of analyte ions. C-5 to C-10 alkylsulfonates are commonly used for cationic compounds; combinations can also be used for difficult separations. Tetraalkyl ammonium salts (tetramethyl-, tetrabutyl-, etc. ammonium salts) or triethyl- (C-5 to C-8) alkyl ammonium salts are generally used in the case of anionic solutes.

The following findings are of some interest in ion-pair separations:

1. Generally, an increase in concentration of pairing reagent increases k' (capacity factor). Ion-pairing reagent concentrations range from 0.001 M to 0.05 M; however, higher concentrations have been endorsed by some researchers.

TABLE 18 Additives for RPLC

Nature of sample	Example	Additive concentration (M or as shown)
Basic compounds	Amines	Phosphate buffer, triethylamine (buffered to pH 3.0)
Acidic compounds	Carboxylic	Phosphate buffer, 1% acetic acid (buffered to pH 3.0)
Mixture of acids and bases	Various	Phosphate buffer, 1% acetic acid (buffered to pH 3.0)
Cationic salts	Tetraalkyl quaternary ammonium compounds	Triethylamine, sodium nitrate
Anionic salts	Alkyl sulfonates	1% acetic acid, sodium nitrate

Note: Adapted from Reference 83a.

2. Greater chain length increases k'. However, when k' is plotted against surface concentration, different chain lengths demonstrate about the same increase in retention. Therefore, most changes of k' seen with increasing chain length may be reproduced with concentration alterations of a single reagent.[84]
3. The k' fluctuates little for neutral species with increases in concentration of the ion-pairing reagent.[84]
4. The k' for solutes that have the same charge as the pairing reagent declines with increases in concentration or chain length of pairing reagents.[84]
5. Elimination of pairing reagents from the column by washing is more troublesome as the chain length increases. This indicates that the use of long-chain pairing reagents can alter the nature of the column.

4. Ion-Exchange Chromatography

Ion-exchange chromatography employs the dynamic interactions between charged solute ions and stationary phases that have oppositely charged groups. In separations of this type, sample ions and ions of like charge in the mobile phase compete for sites (X) on the stationary phase:

$$A^- + X^+B^- \quad B^- + X^-A_- \text{ (anion exchange)}$$
$$C^+ + X^-B^- \quad B^+ + X^-C_+ \text{ (cation exchange)}$$

The magnitude of the ions' competition with B for the charged sites (X) will govern their retention. Generally speaking, this type of chromatography may be used to separate ionic species, such as organic acids or bases, which can be ionized under specific pH conditions. In addition to the reaction with ionic sites on the stationary phase, retention may be affected by the partitioning of solutes between the mobile and stationary phases, as in reversed-phase chromatography. Consequently, even nonionized solutes may be retained on ion-exchange columns.

As discussed previously, ion-exchange chromatography is an adaptable technique used primarily for the separation of ionic or easily ionizable species. The stationary phase is characterized by the presence of charged centers having exchangeable counterions. Both anions and cations can be separated by choosing the suitable ion-exchange medium.[85] Ion exchange is applied in essentially all branches of chemistry. In clinical laboratories, it is used on a routine basis to profile biological fluids and in the diagnosis of various metabolic disorders.[86]

Ion exchange has been used for many years as the separation mechanism in automated amino acid and carbohydrate analyzers. Ion-exchange chromatography was introduced more recently to separate a broad range of biological macromolecules using special wide-pore low-capacity packings that were designed for this purpose.[87–89] Ion exchange may be used to separate neutral molecules as their charged bisulfite or borate complexes, and also certain cations as their negatively charged complexes, for example, $FECl_4^-$. In the case of the borate complexes, carbohydrate compounds that have vicinal diol groups can form stable charged adducts that may be resolved by anion-exchange chromatography.[90] Ligand-exchange chromatography has been used with cation exchangers (in the nickel or copper form) for the separation of amino acids and other bases.

Ion-exchange packings can be used to separate neutral and charged species by mechanisms that do not involve ion exchange. For example, oligosaccharides and related materials can be separated by a partition mechanism on ion-exchange columns where water/alcohol mobile phases are employed.[91] On the basis of their degree of Donnan exclusion from the resin pore volume, ion-exclusion may be used to separate charged species from uncharged species and also charged species from one another. An ion-exchange packing that has the same charge as the sample ions is used for this purpose. Retention depends on the degree of sample access to the packing pore volume. The separation of organic acids with a cation-exchange packing in the hydrogen form is an example of this mechanism.[87,92] Strong acids are totally excluded and elute early; weak acids are somewhat excluded and have intermediate retention values; and neutral molecules are not affected by the Donnan membrane potential and can explore the total pore volume.

The packings for ion-exchange chromatography are typified by the presence of charge-bearing functional groups. Sample retention can be envisioned as a simple exchange between the sample ions and those counterions that were originally attached to the charge-bearing functional groups. However, this simple illustration is not a good portrayal of the actual retention process. Retention in ion-exchange chromatography depends on factors other than coulombic interactions. Hydrophobic interactions between the sample and the nonionic regions of the support are important with organic ions, for example. Since the mobile phase is often of high ionic strengths in ion-exchange chromatography, hydrophobic interactions are favored because of the "salting-out" effect. From a qualitative point of view, the retention of organic ions most likely proceeds by a hydrophobic reversed-phase interaction with the support, followed by diffusion of the sample ion to the fixed charge center where an ionic interaction takes place.

Diffusion-controlled processes influence both column efficiency and sample retention. The following steps are considered important:[59]

- Ion diffusion through the liquid film surrounding the resin bead
- Diffusion of ions within the resin particle to the exchange sites
- Actual exchange of one counterion for another
- Diffusion of the exchanged ions to the surface of the resin bead
- Diffusion of the exchanged ions into the bulk solution through the liquid film surrounding the resin bead

Slow diffusion of the sample ions within the resin beads has a significant detrimental effect on column performance. Reducing the particle size to less than 10 μm in diameter makes up for the poor mass-transfer kinetics exhibited with conventional resin beads by reducing the length of intraparticulate channels.

Because the column packings used in ion exchange contain charged functional groups, an equal distribution of mobile phase ions both inside and outside the resin bead evolves in accordance with the Donnan membrane effects. The ion-exchange bead acts like a concentrated electrolyte solution in which the resin charges are fixed, while the counterions are unrestricted. The contact surface between the resin bead and the mobile phase can be seen as a semipermeable membrane. When equilibration occurs between the external and internal solution and one side of the membrane contains a nondiffusible ion, then a combination of the Donnan membrane effect and the need to preserve overall electrical neutrality results in a higher concentration of free electrolytes within the bead. Thus, diffusion of sample ions and counterions across the Donnan membrane barrier is generally the rate-controlling process in ion-exchange chromatography.

Selectivity series have been established for many counterions:

$$\text{Cations: } Li^+ < H^+ < Na^+ < NH_4^+ < K^+ < Cs^+ < Ag^+ < Cu^{2+}$$
$$< Cd^{2+} < Ni^{2+} < Ca^{2+} < Sr^{2+} < Pb^{2+} < Ba^{2+}$$
$$\text{Anions: } F^- < OH^- < CH_3COO^- < HCOO^- < Cl^- < SCN^-$$
$$< Br^- < I^- < NO_3^- < SO_4^{2-}$$

The absolute order relates to the individual ion exchanger, but deviations from the above order are usually insignificant for different cation and anion exchangers. H^+ is preferred over any common cation for weak acid resins, while OH^- is preferred over any of the common anions weak-base resins.

When a choice of the column type has been made, sample resolution is optimized by adjusting the ionic strength, pH, temperature, flow rate, and concentration of buffer or organic modifier in the mobile phase.[93] A summary of the effects of these parameters on solute retention is given in Table 19.

The temperature at which separations are performed is another variable that can have a marked effect on separations. Temperatures up to 50 °C or 60 °C often result in improved separations because of reduced viscosity and

TABLE 19 Factors Influencing Retention in Ion-Exchange Chromatography (59)

Mobile phase parameter	Influence on mobile phase properties	Effect on sample retention
Ionic strength	Solvent strength	Solvent strength generally increases with an increase in ionic strength. Selectivity is little affected by ionic strength except for samples containing solutes with different valence charges. The nature of mobile-phase counterion controls the strength of the interaction with the stationary phase.
pH	Solvent strength	Retention increases in cation-exchange chromatography and decreases in anion-exchange chromatography with an increase in pH.
	Solvent selectivity	Small changes in pH can have a large influence on separation selectivity.
Temperature	Efficiency	Elevated temperatures increase the rate of solute exchange between the stationary and the mobile phases and also lower the viscosity of the mobile phase.
Flow rate	Efficiency	Flow rates may be slightly lower than in other HPLC methods to maximize resolution and improve mass-transfer kinetics.
Buffer salt	Solvent strength and selectivity	Solvent strength and selectivity are influenced by the nature of the counterion. A change in buffer salt may also change the mobile phase pH.
Organic modifier	Solvent strength	Solvent strength generally increases with the volume percent of organic modifier. Its effect is most important when hydrophobic mechanisms contribute significantly to retention. In this case, a change in the organic modifier can be made to adjust solvent selectivity as normally practiced in reversed-phase chromatography.
Organic modifier	Efficiency	Lowers mobile-phase viscosity and improves solute mass-transfer kinetics.

Note: Adapted from Reference 59.

improved mass transfer. Solute stability at these higher temperatures should be resolved prior to use.

5. Ion Chromatography

There are widespread ion-chromatography applications for the analysis of inorganic and organic ions with pK_a values less than 7. This approach combines the techniques of ion-exchange separation, eluent suppression, and conductivity detection for the quantitative determination of a variety of ions such as mono- and divalent cations and anions, alkylamines, organic acids, and so on.[94–99] The development of the technique is due in part to the difficulty of determining these ions by other methods. Examples of ion chromatographic separations include common anions and the alkali earth elements.[100]

The column packings are styrene–divinylbenzene bead-type resins that are surface-functionalized to confer low ion-exchange capacities of 0.001 to 0.05 M equiv./g.[100, 101] These resins have good structural rigidity, permitting operation at high flow rates with just moderate back-pressure. The mechanical strength of the column packings utilized puts a limit on pressures to approximately 2000 psi. These resin beads are stable over the pH range of 1 to 14. The limited hydrolytic stability of silica-based packings curtails their use in ion chromatography compared to their important position in the modern practice of HPLC. A special packing that has an outer layer of fine (0.1–0.5 μm) aminated latex particles agglomerated to the surface of a surface-sulfonated resin bead is frequently used for anion separations.[102, 103] The latex layer is strongly bound to the surface-sulfonated core by a combination of forces, both electrostatic and van der Waals. The thinness of the exchange layer and the Donnan exclusion effect of the intermediate sulfonated layer together give excellent sample mass-transfer properties by ensuring that single penetration is limited to the outer latex layer.

Generally speaking, the large-sized particles and broad particle-size distributions of the resin packings restrict the efficiency of the columns used in ion chromatography. Resin beads are currently available in the ranges of 20–30, 37–74, and 44–57 μm.

The main problem in the early progress of ion chromatography was the development of an appropriate on-line detection system. Most common ions cannot be detected photometrically, and a technique was required that could detect the separated ions in the background of a highly conducting eluent. Since conductivity is a ubiquitous property of ions in solution and can be simply related to ion concentration, it was determined to be a desirable detection method provided that the contribution from the eluent background could be eliminated. The introduction of eluent suppressor columns for this purpose led to the general acceptance of ion chromatography.[104]

The major disadvantages of suppressor columns are

1. the need to intermittently replace or regenerate the suppressor column
2. the changing elution times for weak acid anions or weak base cations because of ion-exclusion effects in the unexhausted part of the suppressor column
3. the apparent reaction of some ions such as nitrite with the unexhausted part of the suppressor column, with the results of a varying response depending on the percentage exhaustion of the suppressor column;
4. interference in the baseline of the chromatogram by a negative peak typical of the eluent, which varies in elution time with the degree of exhaustion of the suppressor column[59]

5. band spreading in the suppressor column that diminishes the efficiency of the separator column and reduces detection sensitivity

A number of the problems encountered with conventional suppressor columns can be eliminated by using bundles of empty or packed ion-exchange hollow fibers for eluent suppression.[105–107] Because of Donnan exclusion forces, the sample anions in the column eluent do not permeate the fiber wall. The chief advantage of the hollow-fiber ion-exchange suppressor is that it permits continuous operation of the ion chromatography without interferences from baseline discontinuities, ion-exclusion effects, or chemical reactions. The primary disadvantage is that the hollow-fiber ion-exchange suppressors have about 2 to 5 times the void volume of conventional suppressor columns. This, without doubt, leads to some loss of separation efficiency, especially for ions eluting early in the chromatogram. Other limitations include restrictions on utilizable column flow rates to ensure complete suppression and also the necessity for an excess of exchangeable counterions in the regeneration solution.

Common eluents in suppressor ion chromatography are dilute solutions of mineral acids or phenylenediamine salts for cation separations and sodium bicarbonate/sodium carbonate buffers for anion separations. Because these eluents are highly conducting, they cannot be used without a suppressor column or conductivity detection. Fritz and coworkers[108–110] have shown that if separator columns of very low capacity are used in association with an eluent of high affinity of the separator resin, but of low conductivity, a suppressor column is not necessary.

REFERENCES

1. Snyder, L. R. and Kirkland, J. J. *Introduction to Modern Liquid Chromatography*, Wiley, New York, 1979.
2. Ahuja, S., Chase, G. D., and Nikelly, J. G. Pittsburgh Conference on Analytical Chemistry and Spectroscopy, Pittsburgh, PA, March 2, 1964; *Analytical Chemistry*, 37:840, 1965.
3. Ahuja, S. *Selectivity and Detectability Optimizations in HPLC*, Wiley, New York, 1989.
3a. Ahuja, S. *Recent Developments in High Performance Liquid Chromatography*, Metrochem '80, South Fallsburg, New York, October 3, 1980.
4. Ahuja, S. *Trace and Ultratrace Analysis by HPLC*, Wiley, New York, 1992.
5. Ahuja, S. Proceedings of Ninth Australian Symposium on Analytical Chemistry, Sydney, April 27, 1987.
6. Laub, R. J. *Chromatography and Separation Chemistry* (S. Ahuja Ed.), Vol. 297, ACS Symposium Series, p. 1, 1986.
7. Martire, D. and Boehm, R. E. *J. Phys. Chem.* 87, 1045, 1983.
8. McCann, M., Madden, S., Purnell, J. H., and Wellington, C. A. *J. Chromatogr.* 294:349, 1984.
9. Hafkensheid, T. L. *J. Chromatogr. Sci.* 24:307, 1986.
10. Braumann, T., Weber, G., and Grimme, L. H. *J. Chromatogr.* 261:329, 1983.
11. Hafkensheid, T. L. and Tomlinson, E. *Int. J. Pharm.* 16:225, 1983.
12. Nahum, A. and Horvath, C. *J. Chromatogr.* 192:315, 1980.
13. Unger, S. H., Cook, J. R., and Hollenberg, J. S. *J. Pharm. Sci.* 67:1364, 1978.
14. Hoky, J. E. and Michael Young, A. *J. Liq. Chromatogr.* 7:675, 1984.
15. Baker, J. K. *Anal. Chem.* 51, 1693, 1979.
16. Baker, J. K., Skelton, R. E., and Ma, C. Y. *J. Chromatogr.* 168:417, 1979.

17. Chen, B.-K. and Horvath, C. *J. Chromatogr.* 171:15, 1979.
18. Kalizan, R. *J. Chromatogr.* 220:71, 1981.
19. Tomlinson, E. *J. Chromatogr.* 113:1, 1975.
20. Hanai, T. and Jukurogi, A. *Chromatogr.* 19:266, 1984.
21. Hanai, T. and Hubert, J. *J. Chromatogr.* 290:197, 1984.
22. Hanai, T. and Hubert, J. *J. Chromatogr.* 291:81, 1984.
23. Hanai, T. and Hubert, J. *J. Chromatogr.* 302:89, 1984.
24. Melander, W., Campbell, D. E., and Horvath, C. *J. Chromatogr.* 158:215, 1978.
25. Spanjer, M. C. and Deligny, C. L., *Chromatogr.* 20:120, 1985.
26. Jinno, K. and Ozaki, N. *J. Liq. Chromatogr.* 7:877, 1984.
27. Jinno, K. and Kawasaki, K. *Chromatogr.* 18:90, 1984.
28. Arai, Y., Hirukawa, M., and Hanai, T. *J. Chromatogr.* 384:279, 1987.
29. Hanai, T., Jakurogi, A., and Hubert, J. *Chromatogr.* 19:266, 1984.
30. Schoenmakers, P. J., Billiet, H. A. H., and Galan, L. D. *Chromatogr.* 282:107, 1983.
31. Johnson, B. P., Khaledi, M. G., and Dorsey, J. G. *Chromatogr.* 384:221, 1987.
32. Smith, R. M. *Chromatogr.* 324:243, 1985.
33. Cooper, W. T. and Lin, L. Y. *Chromatogr.* 21:335, 1986.
34. Melander, W. R. and Horvath, C. *Chromatogr.* 18:353, 1984.
35. Wang, J. D., Li, J. Z., Bao, M. S., and Lu, P. C. *Chromatogr.* 17:244, 1983.
36. Schoenmakers, P. J., Billiet, H. A. H., and Galan, L. D. *Chromatogr.* 15:205, 1982.
37. Unger, K. K. *Porous Silica*, Elsevier, Amsterdam, 1979.
38. Boumahraz, M., Davydov, V. Y., Gonzalez, M. E., and Kiselev, A. V. *Chromatogr.* 17:143, 1983.
39. Melander, W., Chen, B.-K., and Horvath, C. *J. Chromatogr.* 185:99, 1979.
40. Snyder, L. R. 179:167, 1979.
41. Ulmann, C. L., Genieser, H.-G., and Jastorff, B. ibid., 354:434, 1986.
42. Ahuja, S. "Retention Mechanism Investigations on Ion-pair Reversed phase Chromatography," American Chemical Society Meeting, Miami, April 28, 1985.
43. Ahuja, S. "Probing Separation Mechanism of Ionic Compounds in HPLC," Metrochem '85, Pocono, PA, October 11, 1985.
44. Ahuja, S. "Selectivity and Detectability Optimization in RPLC," Academy of Science USSR Meeting, September 5–7, 1984.
45. Jandera, P., Colin, H., and Guiochon, G. *Anal. Chem.* 54:435, 1982.
46. Colin, H., Guiochon, G., and Jandera, P., through Reference 47.
47. Colin, H., Guiochon, G., and Jandera, P., *Chromatogr.* 17:93, 1983.
48. West, S. D. *J. Chromatogr. Sci.* 25:122, 1987.
49. Snyder, L. R. and Kirkland, J. J. *Introduction to Modern Liquid Chromatography*, p. 261, Wiley, 1979.
50. Snyder, L. R. *J. Chromatogr. Sci.* 16:223, 1974.
51. Lewis, J. J., Rogers, L. B., and Pauls, R. E. *J. Chromatogr.* 264:339, 1983.
52. Smith, R. M. *Anal. Chem.* 56:256, 1984.
53. Goldberg, A. P. *Anal. Chem.* 54:342, 1982.
54. DiCesare, J. L. and Doag, M. W. *Chromatogr. Newsl.* 10, 12–18, 1982.
55. Smith, R. M. *J. Chromatogr.* 236:313, 1982.
56. Smith, R. M., Hurley, T. G., Gill, R., and Moffat, A. C. *Chromatogr.* 19:401, 1984.
57. Sadak, P. C. and Carr, W. *J. Chromatogr.* 21:314, 1983.
58. Verzele, M. and Dewaele, C. *Chromatogr.* 18:84, 1985.
59. Poole, C. F. and Schuette, S. A. *Contemporary Practice of Chromatography*, Elsevier, New York, 1984.
60. Ahuja, S. *Ultratrace Analysis of Pharmaceuticals and Other Compounds of Interest*, Wiley, New York, 1986.
61. Snyder, L. R., Dolan, J. W., and Grant, J. R. *J. Chromatogr.* 165:3, 1979.
62. Simpson, C. F. *Techniques in Liquid Chromatography*, Wiley, New York, 1982.
63. Bristow, P. A. and Knox, J. F. *Chromatographia*, 10:279, 1977.
64. Chen, J.-C. and Weber, S. G. *J. Chromatogr.* 248:434, 1982.
65. Engelhardt, H. *J. Chromatogr.* 15:380, 1977.
66. Thomas, J. P., Burnard, A. P., and Bounine, J. P. *J. Chromatogr.* 172:107, 1979.

67. Engelhardt, H. *High Performance Liquid Chromatography*, Springer-Verlag, Berlin, 1979.
68. Engelhardt, H. and Boehme, W. *J. Chromatogr.* 133:67, 1977.
69. Bredeweg, R. A., Rothman, L. D., and Pfeiffer, C. D. *Anal. Chem.* 51:2061, 1979.
70. Crommen, J. *J. Chromatogr.* 186:705, 1979.
71. Bidlingmeyer, B. A., Del Rios, J. K., and Korpi, J. *Anal. Chem.* 54:442, 198.
72. Smith, R. L. and Pietrzyk, D. G. *J. Anal. Chem.* 56:610, 1984.
73. Majors, R. E. *High Performance Liquid Chromatography, Advances and Perspectives* (Horvath, C., Ed.), Vol. 1, Academic Press, 1980, p. 75.
74. Rassmussen, R. D., Yokoyama, W. H., Blumenthal, S. G., Bergstrom, D. E., and Ruebner, B. H. *Anal. Biochem.* 101:66, 1980.
75. Wise, S. A., Chesler, S. N., Hertz, H. S., Hilpert, L. P., and May, W. E. *Anal. Chem.* 49:2306, 1977.
76. Chmielowiec, J. and George, A. E. *J. Anal. Chem.* 52:1154, 1980.
77. Karlesky, D., Shelley, D. C., and Warner, I. *J. Anal. Chem.* 53:2146, 1981.
78. Krstulovic, A. M. and Brown, P. R. *Reversed-Phase High Performance Liquid Chromatography: Theory, Practice and Biomedical Applications*, Wiley, New York, 1982.
79. Jandera, P., Colin, H., and Guiochon, G. *Anal. Chem.* 54:435, 1982.
80. Otto, M. and Wegscheider, W. *J. Chromatogr.* 258:11, 1983.
81. Melander, W. R. and Horvath, C. *High Performance Liquid Chromatography, Advances and Perspectives*, Vol. 2, Academic Press, New York, 1980, p. 114.
82. Scott, R. P. W. and Kucera, P. *J. Chromatogr.* 142:213, 1977.
83. Colin, H. and Guiochon, G. *J. Chromatogr.* 158:183, 1978.
83a. Snyder, L. R., Glajch, J. L., and Kirkland, J. J. *Practical HPLC Method Development*, Wiley, New York, 1988.
84. Knox, J. A. and Harwick, R. A. *J. Chromatogr.* 204:3, 1981.
85. Elchuk, S. and Cassidy, R. M. *Anal. Chem.* 51:1434, 1979.
86. Miyagi, H., Miura, J., Takata, Y., Kamitake, S., Ganno, S., and Yamagata, Y. *J. Chromatogr.* 239, 733, 1982.
87. Regnier, F. E., Gooding, K. M., and Chang, S.-H. *Contemporary Topics in Analytical Clinical Chemistry*, Vol. 1, Plenum, New York, P. 1, 1977.
88. Regnier, F. E. and Gooding, K. M. *Anal. Biochem*, 103:1, 1980.
89. Vacik, D. N. and Toren, E. C. *J. Chromatogr.* 228:1, 1982.
90. Voelter, W. and Bauer, H. *J. Chromatogr.* 126:693, 1976.
91. Samuelson, O. *Adv. Chromatogr.* 16:113, 1978, through Reference 2.
92. Tanaka, K. and Shizuka, T. *J. Chromatogr.* 174:157, 1979.
93. Rabel, F. M. *Adv. Chromatogr.* 17:53, 1979, through Reference 59.
94. Sawicki, E., Mulik, J. D., and Wattgenstein, E. *Ion Chromatographic Analysis of Environmental Pollutants*, Vol. 1, Ann Arbor, 1978.
95. Mulik, J. D. and Sawicki, E. *Ion Chromatographic Analysis of Environmental Pollutants*, Vol. 2, Ann Anbor, 1979.
96. Smith, F. C. and Change, R. C. *CRC Crit. Revs. Anal. Chem.* 9:197, 1980.
97. Small, H. *Trace Analysis*, Vol. 1 (Lawrence, J. F. Ed.), Academic Press, New York, 1982, p. 269.
98. Fritz, J. S., Gjerde, D. T. and Pohlandt, C. *Ion Chromatography*, Huthig, Heidelberg, 1982.
99. Small, H. *Anal. Chem.* 55:235A, 1983.
100. Pohl, C. A. and Johnson, E. L. *J. Chromatogr. Sci.* 18:442, 1980.
101. Gjerde, D. J. and Fritz, S. J. *J. Chromatogr.* 176:199, 1979.
102. Stevens, T. S. and Small, H. *J. Liq. Chromatogr.* 1:123, 1978.
103. Stevens, T. S. and Langhorst, M. A. *Anal. Chem.* 54:950, 1982.
104. Small, R., Stevens, T. S., and Bauman, W. C. *Anal. Chem.* 47:1801, 1981.
105. Stevens, T. S., Davis, J. C., and Small, H. *Anal. Chem.* 53:1488, 1981.
106. Hanaoki, Y., Murayama, T., Muramoto, S., Matsura, T., and Nanba, A. *J. Chromatogr.* 239:537, 1982.
107. Stevens, T. S., Jewett, G. L., and Bredeweg, R. A. *Anal. Chem.* 54:1206, 1982.
108. Gjerde, D. T., Fritz, J. S., and Schmuckler, G. S. *J. Chromatogr.* 186:509, 1979.
109. Gjerde, D. T., Fritz, J. S., and Schmuckler, G. S. *J. Chromatogr.* 187:35, 1980.
110. Matsushita, S., Tada, Y., Baba, N., and Hosako, K. *J. Chromatogr.* 259:459, 1983.

QUESTIONS FOR REVIEW

1. List various modes of chromatography. Which mode of chromatography is most commonly used in HPLC?
2. How can you vary α values in RPLC?
3. Describe various approaches to solvent optimization in RPLC.
4. List some of the common additives for RPLC.

11
EVOLVING METHODS AND METHOD SELECTION

I. EVOLVING METHODOLOGIES
   A. Supercritical Fluid Chromatography
   B. Field Flow Fractionation
   C. Electrophoretic Methods
   D. Capillary Electrochromatography
II. SELECTION OF A SEPARATION METHOD
III. CHIRAL SEPARATIONS
   A. Thin-Layer Chromatography
   B. Gas Chromatography
   C. High-Pressure Liquid Chromatography
   D. Supercritical Fluid Chromatography
   E. Capillary Electrophoresis
   F. Membrane Separations
   G. Enzymatic Methods
IV. COMPARISON OF GC, SFC, HPLC, AND CEC FOR A SELECTIVE SEPARATION
   REFERENCES
   QUESTIONS FOR REVIEW

A number of separation methods have been discussed in the earlier chapters, which include versatile, efficient, and highly useful chromatographic methods. It is important to recognize that a method useful for a given separation can be as simple as filtration or as complex as some newly evolved separation methods discussed later in this chapter (see Section I). The fact is that new and more efficient methods are constantly being developed to provide better resolution in a shorter period of time and to yield maximum information on a progressively smaller sized sample.

It is a fair guess that in ancient times mankind most likely relied mainly on sedimentation to yield clear liquids. This means a long waiting period before a clear liquid could be decanted. Indubitably, filtration through some medium was the fastest approach one could use to achieve the desired goal. The main point to note here is that earlier attempts to achieve clear liquids were focused primarily on separating visible particles. The problems stem from the fact that particles come in various sizes; some are so small that they are not visible to human eye without magnification. It is important to realize that there is no sharp discontinuity between very large molecules, colloid particles, and macroscopic particles.

With due consideration to the above comments, separation methods can be broadly divided into the following two classes:

1. Separations of particulated samples
2. Separations of nonparticulated samples

It is meaningful to recognize that this classification is somewhat arbitrary. We will learn later in this chapter that some of the separation methods, such as field flow fractionation, can separate particles as well as macromolecules.

Separations of Particulated Samples: Filtration methods can provide a broad range of separations that are able to separate visible particles with simple filter papers. Utilizing sophisticated membranes, it is possible to filter out small particles and bacteria that are invisible to the human eye and can be seen only with a microscope. Let us review the methods that can be used for separation of particles (the alphabetical list is repeated here from Chapter 1).

- Electrostatic precipitation
- Elutriation
- Field flow fractionation
- Filtration
- Flotation
- Particle electrophoresis
- Precipitation
- Screening
- Sedimentation
- Ultracentrifugation

Of these methods, field flow fractionation, which can be related to equilibrium sedimentation, is a relatively recent method and is discussed in Section I. As mentioned before, the simplest form of separation entails removal of particles that are discernible. Filtration, screening, and sedimentation have been used for this purpose from time immemorial. These processes are generally referred to as mechanical processes. A well-known example of mechanical separation processes is elutriation, where particles are separated according to their size and shape.

Centrifugation and ultracentrifugation are convenient methods that achieve separation of various sized particles by enhancing the sedimentation rate of particles with the help of centrifugal force. Ultracentrifugation is more useful for smaller particles that require greater centrifugal force to achieve the desired results.

Precipitation generally entails a chemical reaction to precipitate out the desired material. However, physical phenomena such as salting out, where a solution of the desired material is saturated with certain salts, can be used to precipitate the desired material. Of course, it may sometimes be more desirable to precipitate the undesired material out instead, thereby leaving the desired material in the filtrate. Electrostatic precipitation makes use of the electrical charge on materials to achieve this goal; neutralization of the charge generally leads to the desired results.

Particle electrophoresis is a separation method based on the differential rate of migration of charged species in a buffer solution across which a direct-current electric field has been applied. This has led to the modern techniques such as capillary electrophoresis discussed in Section I.

Separations of Nonparticulated Samples: It should be readily apparent from the previous discussion that separation methods become more complex as the size of particles decreases. As mentioned above, there is no sharp discontinuity between macroscopic particles, colloidal particles, and very large molecules. This complexity increases even further when a molecule is solubilized in a solvent; that is, it is no more a defined particle in the conventional sense. The methods used for the separations of solubilized materials have been covered at length in this book.

A special mention needs to be made of the wide range of separations that are possible by utilizing TLC, GC, and HPLC methods. Thin-layer chromatography (see Chapter 8) provides an inexpensive way of separating a large variety of compounds for qualitative, semiquantitative, or quantitative analysis. Sophisticated methods such as gas chromatography and various modes of HPLC (discussed in Chapters 9 and 10) are especially useful for a large number of separations of volatile, nonvolatile, low and high molecular weight compounds.

This chapter deals with methods that have evolved relatively recently (see Section I) and describes strategies for optimum selection of a separation method (see Section II). The reader may also want to consult a number of other useful texts that are available in this area.[1-5] A special section on chiral separations (Section III) is included in this chapter to highlight the uniqueness of the separation of compounds that differ only as a molecule and its mirror image.

I. EVOLVING METHODOLOGIES

Various evolutionary paths led to the development of a variety of chromatographic methods (discussed earlier in this book). In most analytical-scale separation methods, the sample components are distributed over the separation path as distinct zones. The containment of the zones is a major goal of separation scientists. Most forms of chromatography and electrophoresis illustrate this. In these cases, the zones spread continuously outward as the separation process advances. However, the successful separation depends on keeping the zones reasonably narrow to avoid overlap and contamination with neighboring zones. The basic laws controlling the structure and evolution of zones and various indices such as plate height, number of plates, peak capacity, and resolution that measure the effectiveness of separation were discussed earlier in this book.

A number of chromatographic and nonchromatographic methods have been developed relatively recently or have evolved to yield new applications. Some of these methods are listed here:

- Supercritical fluid chromatography
- Field flow fractionation

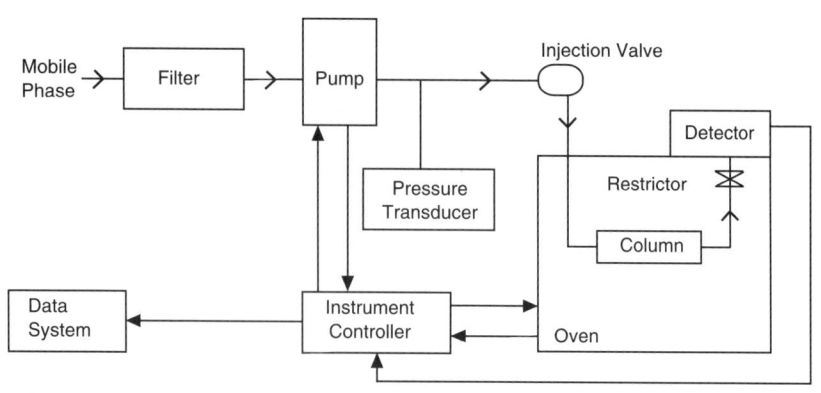

FIGURE I Block diagram of an SFC instrument.

- Electrophoretic methods
- Capillary electrochromatography

A. Supercritical Fluid Chromatography

Supercritical fluid chromatography (SFC) is a hybrid technique of gas chromatography (GC) and HPLC that combines some of the best features of both methods. In 1985, several instrument manufacturers began to offer equipment specifically designed for SFC. A schematic of the apparatus used is shown in Figure 1. The instrument development for SFC arose out of what had been already developed for GC and HPLC. The supercritical fluid delivery system is basically a pump modified for pressure control, and the injection system utilizes a rotary valve similar to that used in HPLC. The column oven and flame ionization detector are similar to those used in GC. Detectors used in HPLC can be also used with appropriate modification for higher pressure operation. An important requirement in SFC is a restrictor for maintaining constant density of the mobile phase along the column length and maintaining the flow rate through the column. A number of commercial SFC instruments are available today.

In SFC the mobile phase has a low viscosity and a high self-diffusion coefficient. This is achieved by utilizing various gases that produce supercritical fluids of low density at critical temperature and pressure.

Gas	Temp. (°C)	Pressure (atm.)	Density (g/ml)
Carbon dioxide	31.3	72.9	0.47
Nitrous oxide	36.5	72.5	0.45
Ammonia	132.5	112.5	0.24

Of these gases, carbon dioxide is most commonly used.

The mobile phase can be further modified by addition of polar modifiers in small amounts.

II EVOLVING METHODS AND METHOD SELECTION 213

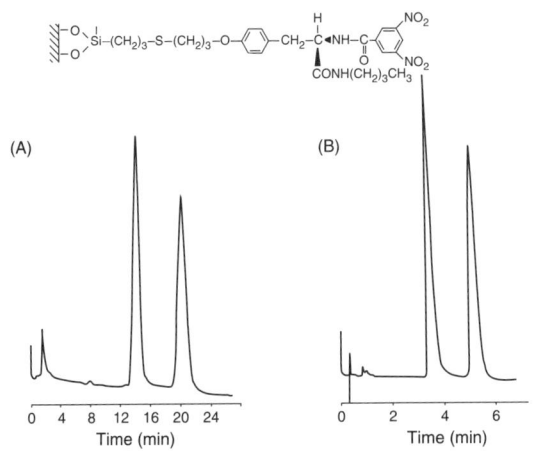

FIGURE 2 Separation of oxazepam by HPLC and subFC.[6]

I. Advantages of SFC over GC and HPLC

SFC has gained importance as a separation technique because it allows separation of compounds that are not conveniently separated by GC or HPLC. It offers a number of advantages of both GC and HPLC.

1. The ability to program the density or pressure provides a unique advantage in SFC over GC and HPLC in that it allows the chromatographer to increase the solvating power of the solvent by increasing the density.
2. The mobile phase selectivity offered by HPLC can still be brought into play by using various additives.
3. The mobile phase can be easily volatilized as compared to HPLC.
4. An advantage over HPLC is that detectors employed in GC can be utilized in SFC.

At times, subcritical fluid chromatography (subFC) can be used to advantage. Figure 2 shows separations for oxazepam by HPLC and subFC on the same CSP. It may be noted that separation time is shorter by subFC, and peak width is narrower than with HPLC.

B. Field Flow Fractionation

Field flow fractionation(FFF) is carried out in a thin (50–500 µm) ribbonlike flow channel, with an external field or gradient applied across the channel faces (see Fig. 3). The field serves to force molecules or particles toward one wall, where their downstream velocity is diminished because of the sluggish flow of carrier fluid caused by the drag on the wall. Components that interact strongly are pushed into layers against the wall, which may extend no more than 1–10 µm into the flow stream.[7]

The component molecules or particles soon establish a thin steady-state distribution in which outward diffusion balances the steady inward drift due to the field. The structure and dimension of this layer determine the separation

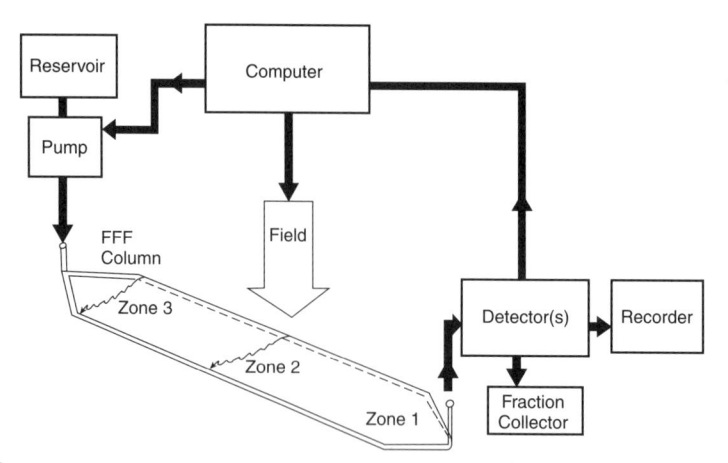

FIGURE 3 A schematic of field flow fractionation.[7]

process. Similar layers can be formed at the outside wall of a centrifuge tube when the centrifugation process is allowed to proceed to equilibrium.

The range of molecular weight of compounds that can be analyzed by FFF as compared to other separation methods is shown in Figure 4. We can see from the figure that GC, HPLC, and SEC are limited to molecular weights around 1 million and particle sizes ranging from 0.0001 μm to 0.01 μm, whereas, FFF can handle molecular weights up to 10^{18} molecular weight and particle sizes as high as 100 μm.

A number of core methods are available in FFF. The comparative diversity of chromatography to FFF is shown in Table 1. Thermal FFF, for example, allows separation of a nine-component mixture of polystyrene in ethylbenzene (Fig. 5). After 1 h at constant field characterized by a hot wall temperature of 87 °C, the field is reduced to zero over a 6-h period using a parabolic decay program. The cold wall is maintained at 27 °C.

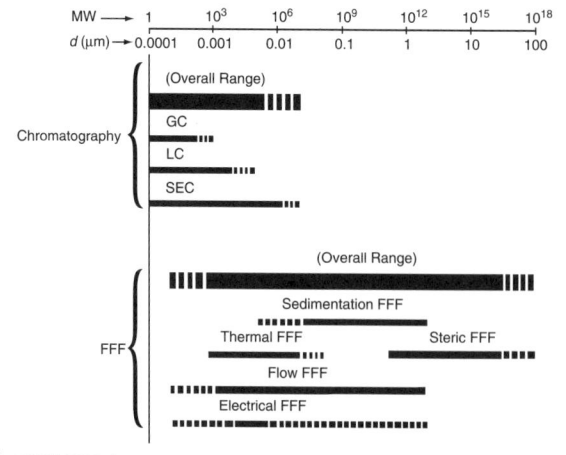

FIGURE 4 Comparative range of separations with chromatography and FFF.[7]

TABLE I Comparison of Chromatography and FFF

	Chromatography	FFF
Core Methods	GLC, GSC, LLC, LSC, IEC, reversed-phase LC	Thermal FFF, electrical FFF, sedimentation FFF, magnetic FFF, FFF with other fields
Major Variants:		
1. Altered flow	TLC	Flow FFF(solvent exchange), thermogravitational FFF
2. Altered basis of retention	Size-exclusion chromatography	Steric FFF
3. Program variables	Temperature, flow, solvent strength	Field strength, flow, carrier fluid properties
4. Other	Continuous operation, mixed stationary phases	Continuous operation, combined fields

Note: Adapted from Reference 7.

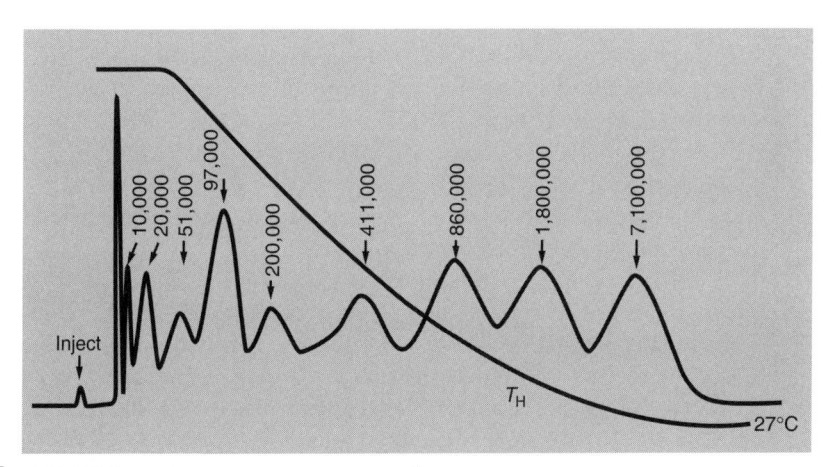

FIGURE 5 Thermal FFF of polystyrenes.[7]

DNA mixtures have been separated by FFF. Both single-stranded and double-stranded molecules can be resolved. Globular proteins—cytochrome C, albumin, and thyroglobulin—have also been resolved with flow FFF.

C. Electrophoretic Methods

Electrophoresis is defined as transport of electrically charged particles in a direct-current electric field. The particles may be simple ions or complex macromolecules and colloids, or they may be particulate matter such as living cells (bacteria or erythrocytes) or inert material (oil emulsion droplets or clay). Electrophoretic separation is based on differential rate migration in the bulk of the liquid phase, and it is not concerned with reactions occurring at the electrodes.

The Swedish chemist Arne Tiselius was the first to develop electrophoresis in the 1930s for the study of serum proteins. The Nobel Prize in chemistry was awarded to him in 1948 for his work. Electrophoresis can be performed in two different formats:

- Electrophoresis on a slab of gel
- Electrophoresis in a capillary

I. Electrophoresis on a Slab of Gel

Slab electrophoresis is a classical method that is carried out on a thin flat layer or slab of a porous semisolid gel containing aqueous buffer solution in its pores. The slab has dimensions of a few centimeters on a slide, and like a TLC plate, it is capable of separating samples simultaneously. Samples are introduced as spots or bands and direct current potential is applied across the slab for a fixed period of time. When the separations are considered finalized, the current is discontinued and the separated species are visualized by staining.

2. Electrophoresis in a Capillary

Electrophoresis carried out in a capillary is called capillary electrophoresis. Capillary electrophoresis (CE) is the instrumental version of electrophoresis that is carried out in a capillary; it emulates chromatographic techniques in terms of simplicity of operation. However, as discussed below, CE is not strictly a chromatographic technique. The main advantages of CE are high-speed, high-resolution separations on exceptionally small volumes. The separated species of interest come out at the end of a capillary, where detectors similar to HPLC can be used for detection and quantitation.

Strictly speaking, CE is not a chromatographic technique because two phases are not involved in the separation process. The two phases in chromatography are designated as the stationary phase and the mobile phase, based on their role in the separation process, and technically, there is no stationary phase in CE unless the capillary walls are assigned that role. Some researchers encourage this concept, but it is not valid. In any event, most chromatographers are comfortable in using CE because the technique has a number of similarities to chromatography in that some of the manipulations that are employed to optimize chromatographic separations are also suitable for CE. Moreover, major chromatographic meetings often include symposia on CE.

Capillary electrochromatography, discussed later in Section I. D, is indeed a chromatographic technique that is a hybrid of electrophoresis and liquid chromatography.

3. Capillary Zone Electrophoresis

Broadly speaking, electrophoresis can be classified on the basis of whether it is carried out as a free solution or on the support media. When a support medium is used, the technique is called zone electrophoresis. Capillary electrophoresis that is commonly carried out today fits into this category, and at one time was called capillary zone electrophoresis (CZE). The four major modes of capillary electrophoresis are given in Table 2.

TABLE 2 Modes of Capillary Electrophoresis

Mode	Separation mechanism
Free solution	Based on solute size and charge at a given pH
SDS Page	Based on solute size
Isoelectric focusing	Based on solute isoelectric point
MECC	Based on solute partitioning between micelle "phase" and solution phase

SDS-Page: Sodium dodecyl sulfate polyacrylamide gel electrophoresis.
MECC: Micellar electrokinetic capillary chromatography. MECC is at times called MEKC.

TABLE 3 High-Performance Capillary Electrophoresis

Name	Basis of separation	Contributor
Free-zone CE	Electrophoretic mobility	a
MECC	Partitioning into detergent micelles, charge	b
Isoelectric focusing	Isoelectric point	c
Chiral separation	Enantiomeric structure	d
Gel electrophoresis: polyacrylamide	Size, charge	e
SDS-Page	Size	f

MECC = Micellar electrokinetic capillary chromatography; also called MEKC.
a. Jorgensen and Lukacs, *Anal. Chem.* 53:1298 (1981).
b. Terabe *et al.*, ibid, 56:113 (1984).
c. Hjerten and Zhu, *J. Chromatogr.* 346:265 (1985).
d. Gassman *et al.*, *Science* 242:813, (1985).
e. Cohen *et al.*, *Nat. Acad. Sci.* 85:9660 (1988).
f. Cohen and Karger, *J. Chromatogr.* 397:409 (1987)

Table 3 includes chiral separations, original contributor(s) to this field, and the basis of separation in high-performance capillary electrophoresis.

It is important to remember that highest resolution is obtained when an element of discontinuity is introduced in CE in the liquid phase, such as pH gradient, or the sieving effect of high-density gels. Membrane barriers may also be introduced into the path of migrating components.

In capillary zone electrophoresis, the buffer composition is constant throughout the region of separation. The applied potential causes different ionic components to migrate according to their own mobility and to separate into zones that may be completely resolved or may partially overlap. This is analogous to elution column chromatography. This is the method of choice for most electrophoretic separations of small ions. A variety of pharmaceuticals and synthetic herbicides and pesticides that are ionic or can be derivatized to yield ions can be separated and analyzed by CZE. Proteins and amino acids are ideal candidates for this technique. Neutral carbohydrates require formation of negatively charged borate complex prior to analysis with this technique.

4. Capillary Gel Electrophoresis

Capillary gel electrophoresis (CGE) is generally performed in a porous polymer matrix, the pores of which contain a buffer mixture in which the separation is carried out. This type of medium can provide molecular sieving action that retards the migration of analytical species to various extents, depending on the pore size of the polymer and the size of analyte ions. This technique is useful for separating macromolecules such as proteins, DNA fragments, and oligonucleotides that differ in size but have substantially the same charge.

5. Micellar Electrokinetic Capillary Chromatography

Micellar electrokinetic capillary chromatography (MECC) utilizes partition in detergent to emulate chromatography.

6. Capillary Isotachophoresis

Isotachophoresis derives its name from *iso* (meaning same) and *tach* (meaning speed). In capillary isotachophoresis, all analyte bands migrate at the same velocity. In any particular application, either cations or anions can be separated. It is not possible to separate both of them simultaneously. To carry out a separation, the sample is injected between two buffers, the leading buffer containing ions of higher mobility and the terminating buffer with ions of lower mobility than the sample ions.

Capillary isoelectric focusing is used to separate amphiprotic species such as amino acids and proteins that contain a weakly acidic carboxylic acid group and a weakly basic amine group.

7. Schematic of a CE Instrument

In CE a relatively short capillary, attached to the respective reservoirs, is subjected to high voltages, around 30 kV (Fig. 6). The capillary tube has a diameter of 75 μm or less, which allows an easy dissipation of generated heat through the wall. The sample can be drawn inside the capillary tube by a short exposure to high voltage. The zone breadth is proportional to the applied voltage.

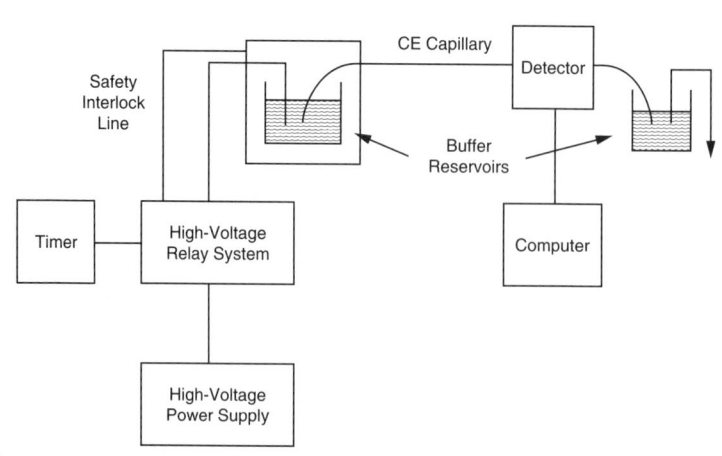

FIGURE 6 Block diagram of a CE instrument.

TABLE 4 Comparison of Modes of Separations of CE and HPLC

HPLC	CE
Mobile phase additive	FSCE with crown ether
Cyclodextrins	FSCE with cyclodextrins
	MECC with cyclodextrins
Proteins	FSCE with protein additives
Carbohydrate phase	FSCE with sugar additive
Chiral ligand exchange	FSCE with enantioselective chelation
	MECC with enantioselective chelation

8. Comparison of CE to HPLC

A comparison of modes of separations used in CE and HPLC is given in Table 4. Capillary electrophoresis has been found to be quite useful for resolving a very large number of compounds.

The inherent simplicity of CE has attracted many researchers to utilize it for chiral separations. It offers great potential that is sure to result in many useful applications in the future. Two modes of CE that are commonly used are free solution capillary electrophoresis (FSCE) and micellar electrokinetic capillary chromatography (MECC). CE is a very useful technique for enantiomeric separations (see Section III). The primary advantage of CE is that it can offer rapid, high-resolution of water-soluble components present in small volumes. The separations are based, in general, on the principles of electrically driven flow of ions in solution. Selectivity is accomplished by alteration of electrolyte properties such as pH, ionic strength, and electrolyte composition, or by the incorporation of electrolyte additives. Some of the typical additives include organic solvents, surfactants, and complexing agents.

The degree of chiral recognition attainable in chiral CE is often similar to that obtained with HPLC. However, the high efficiency offered by CE as compared with HPLC offers a significant advantage. A major advantage of CE is that it is possible to analyze a small sample. Of course, this proves to be a limitation if a larger sample size is desirable, as for example, in preparative separations.

D. Capillary Electrochromatography

The hybrid technique of capillary electrochromatography (CEC) combines the separation power of reversed-phase HPLC and the high efficiency of capillary electrophoresis. In CEC, separation of an uncharged molecule is achieved on the basis of differential partitioning into the stationary phase. The mobile phase is pumped electrically; as a result, the analytes are carried through the column by electroosmotic flow. Basically, CEC differs from CE in that the capillary is packed with stationary-phase particles and requires a suitable frit to contain the packing material. Because there is a lack of pressure limitations in CEC, the stationary-phase particles can be reduced theoretically to a submicrometer level.

This should minimize two sources of band broadening in chromatography, namely, eddy diffusion and resistance to mass transfer.

This technique has been successfully used for a wide range of compounds, which includes polycyclic aromatic hydrocarbons and pharmaceutical compounds, together with chiral separations. It should be pointed out that the majority of available publications make use of 3-μm particle size C-18 stationary phase.

II. SELECTION OF A SEPARATION METHOD

The best place to start is to find out what information is currently available that relates to separation of the compounds of interest. This generally means doing a literature search or talking to a knowledgeable scientist. Not only will this approach lead to a better selection of methods and minimize loss of time on trial-and-error methods, it will also eliminate the costly reinvention of the wheel. Following the initial search, planned logical experiments should be carried out to assure optimum selection of the method.

A complete search might start with *Chemical Abstracts* or *Analytical Abstracts*. Computer-based data searches are now possible, and the Internet has made this operation more rapid and accessible. If the choice is limited to a single method such as GC or HPLC, select literature surveys are available in this area.[8–11] A quick review of compilations of retention data by ASTM for GC, HPLC, and SEC may be helpful in getting started.[12–14] Recent information can be found in a review of bibliographic data in *Analytical Chemistry*. The *Journal of Chromatography* also publishes bibliographic data in the following areas:

- Electrophoresis
- GC
- HPLC
- Planar chromatography

A number of criteria can be used in the selection of a separation method. These criteria help classify the major separation methods (Table 5). They are presented in the order of increasing importance. Categories 1–4 of Table 5 indicate the sample type that can be used with a given method. And categories 5–8 show the type of method that is being used. Designations *a* or *b* indicates that this method is limited to this particular sample and is most commonly used for this purpose. The designation *a,b* indicates that both sample types can be used. A review of the table will reveal that no two methods fit exactly the same classification, although several are quite close. The table can help make a decision in the preliminary selection of a separation method.

A review of this table also reveals that simple extraction methods can be used for hydrophilic, hydrophobic, ionic, nonionic, or nonvolatile samples. For volatile samples, distillation or gas chromatography may be the method of choice. It goes without saying that simple and quick methods should be

II EVOLVING METHODS AND METHOD SELECTION

TABLE 5 Usefulness of Various Separation Methods

Method	Sample type*	Method type*
Extraction	1a,b;2a,b;3b;4a	5a;6a,b;7a,b;8a,b
Distillation	1b;2b;3a;4a	5b;6a;7a;8b
Adsorption chromatography	1b;2b;3b;4b	5a;6a,b;7b;8a,b
HPLC	1a,b;2a,b;3b;4b	5a;6a;7b;8a,b
TLC	1a,b;2a,b;3b;4a	5b;6b;7b;8a,b
GC	1b;2b;3a;4b	5a;6a;7b;8a
Ion-exchange chromatography	1a;2a;3b;4b	5a;6a;7b;8a,b
Size-exclusion chromatography	1a,b;2b;3b;4b	5a;6a,b;7b;8a,b
Capillary electrophoresis	1a;2a;3b;4a	5b;6a,b;7b;8a

Adapted from Miller, J. M. *Separation Methods in Chemical Analysis*, Wiley, NY, 1975.
*Key:
1. a = hydrophilic; b = hydrophobic
2. a = ionic; b = nonionic
3. a = volatile; b = nonvolatile
4. a = simple; b = complex
5. a = quantitative; b = qualitative
6. a = individual; b = group type
7. a = recovery; b = purity
8. a = analytical; b = preparative

the preferable choices if they provide the desired selectivity. More often than not, chromatographic methods would be indicated because of the speed and selectivity offered by them. If a preliminary review reveals that a chromatographic method is the method of choice, there are three major objectives that need to be considered:[15]

- Resolution
- Quantity of sample to be analyzed
- Speed of analysis

These objectives are interrelated. Based on some empirical judgments, it can be said that TLC and gravity-fed column liquid chromatography are primarily useful for processing a large quantity of sample.[3] The speed or resolution offered by these methods is rather low. GC is the technique of choice when speed and resolution are the primary considerations. HPLC is useful when the quantity of sample to be processed is large; however, the resolution of several components is an important consideration.

Table 6 gives the characteristics of individual methods that may be considered in the selection process. TLC is an excellent technique for preliminary screening and for semiquantitative evaluations of samples. Quantitation is possible; however, it can be cumbersome, and precision demands great care. SEC can be used for a review of molecular weights of the components of the sample. Clearly, different approaches are needed for handling low molecular weight samples (< 2000) than are needed for macromolecules (e.g., see Reference 16).

TABLE 6 Characteristics of Various Separation Methods

Method	Sample types	Separation basis	Fractional capacity	Load capacity	Speed
Extraction	Nonvolatile; large molecules	Physicochemical	2	g–kg	Adequate
Distillation	Volatile	Mol. Weight	10	g–kg	Fair
GC	Volatile	Mol. Weight	>100	mg–g	Good
Dialysis	Nonvolatile	Mol. Weight	2	mg–g	Fair
Ultrafiltration	Nonvolatile; large molecules	Mol. Weight	2	mg–kg	Fair
Paper chromatography	Nonvolatile	Physicochemical	10	µg–mg	Slow
TLC	Nonvolatile	Physical	10–50	mg–g	Good
RPLC	Nonvolatile	Physicochemical	10–50	ng–mg	Good
CE	Nonvolatile	Physicochemical	>10	ng–µg	Good
SEC	Nonvolatile	Mol. Weight	10	mg–g	Good
Ion exchange	Nonvolatile	Ions	10–100	mg–g	Good
Ultracentrifugation	Large molecules	Mol. Weight	5	mg	Adequate

Adapted from Reference 1.

Figure 7 provides a selection process of various HPLC methods based on the following sample properties:

- Molecular weight
- Solubility in organic or aqueous solvent
- Sample type, e.g., polar, ionic, etc.

Polymeric materials are frequently resolved by gel-permeation or size-exclusion chromatography, based on their solubility in water or organic solvents. At times, elaborate schemes have to be designed prior to the analysis of

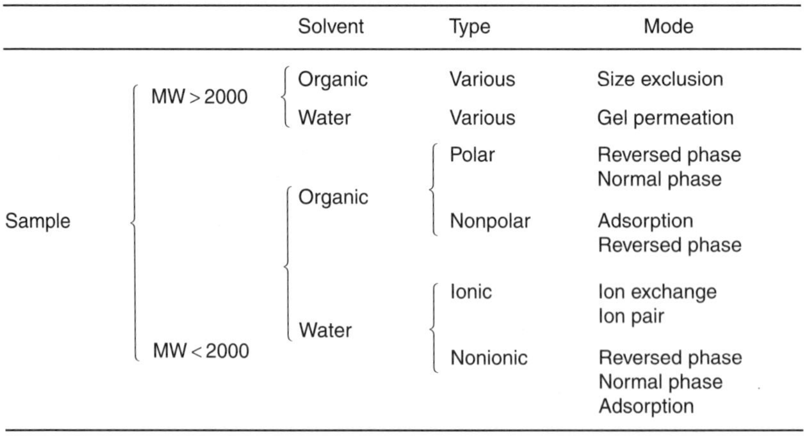

	Solvent	Type	Mode
MW > 2000	Organic	Various	Size exclusion
	Water	Various	Gel permeation
MW < 2000	Organic	Polar	Reversed phase / Normal phase
		Nonpolar	Adsorption / Reversed phase
	Water	Ionic	Ion exchange / Ion pair
		Nonionic	Reversed phase / Normal phase / Adsorption

FIGURE 7 Choice of method based on sample properties.

II EVOLVING METHODS AND METHOD SELECTION

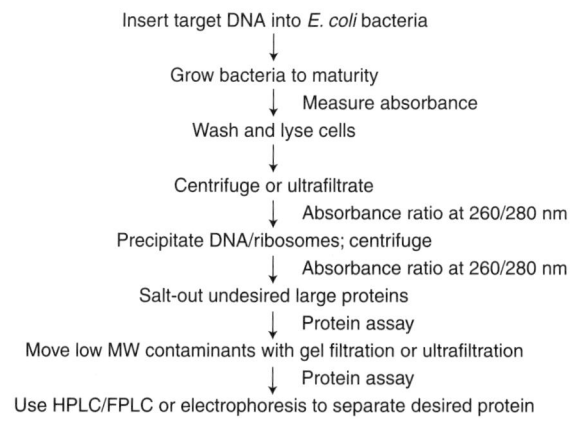

```
              Insert target DNA into E. coli bacteria
                              ↓
                   Grow bacteria to maturity
                              ↓   Measure absorbance
                    Wash and lyse cells
                              ↓
                    Centrifuge or ultrafiltrate
                              ↓   Absorbance ratio at 260/280 nm
              Precipitate DNA/ribosomes; centrifuge
                              ↓   Absorbance ratio at 260/280 nm
                 Salt-out undesired large proteins
                              ↓   Protein assay
       Move low MW contaminants with gel filtration or ultrafiltration
                              ↓   Protein assay
          Use HPLC/FPLC or electrophoresis to separate desired protein
```

FIGURE 8 Typical purification procedure in biotechnology.

the sample. An example of a typical purification procedure in a biotechnology laboratory is given in Figure 8. It should be noted that a variety of separation/chromatographic procedures are being used in the scheme illustrated. This suggests the need of personnel well trained in these techniques, coupled with adequate equipment. It is very desirable that similar extraction/purification schemes be designed before initiating a complex separation problem.

III. CHIRAL SEPARATIONS

Chiral separations utilize a variety of methods covered in this book. The methods used for chiral separations are included in one section here because they rely on evolving methodology that provides unique selectivity for molecules that have the same molecular structure but different stereomeric configuration, which results in different optical rotations (dextro or levo) for the enantiomers. Since these methods embrace a variety of methodologies and techniques required to provide an optimum separation for molecules that have the same physicochemical properties, it is envisioned that the detailed discussion given below will help with a better appreciation of separation methods. The frequently used techniques for separation of enantiomers can be broadly classified into the following seven categories:

- Thin-layer chromatography (TLC)
- Gas chromatography (GC)
- High-pressure liquid chromatography (HPLC)
- Supercritical fluid chromatography (SFC)
- Capillary electrophoresis (CE)
- Enzymatic methods
- Membrane separations

It should be apparent that the last three methods are not based on chromatography, although chromatography is extensively used in various steps of enzymatic separations. The discussion that follows is limited to a short introduction of these techniques and how they are best utilized for enantiomeric separations.

A. Thin-Layer Chromatography

As noted in Chapter 8, TLC is a very useful qualitative method that involves minimal costs. It can afford good indications as to which technique would be most appropriate for resolving enantiomers. It can, of course, also be used as an independent technique with the well-known limitations of resolution and low precision. A significant amount of coverage is given here to encourage the reader to try TLC. Additional reference sources are also provided for TLC devotees.

As mentioned before, in all chromatographic methods including TLC, two phases are essential to accomplish a successful separation. These phases are specified as the mobile phase and the stationary phase. In conventional TLC, the stationary phase is generally silica gel, and the mobile phase is composed of a mixture of solvents. Contrary to common notions, the perceived separation in TLC is not by virtue of adsorption on silica gel alone. There is always a finite quantity of water existing in the silica gel plates; this acts as a partitioning agent. Moreover, the mobile phase solvents are adsorbed onto silica during development and therefore afford yet another mechanism for partition. Ionic sites in silica gel permit ion exchange, and metallic impurities provide mixed mechanisms for specific separations.

It may be recalled that the approaches to separation of chiral chromatography can be classified as follows:

- Achiral stationary phases and achiral mobile phases
- Achiral stationary phases and chiral stationary phase additives
- Chiral stationary phases (CSPs) and achiral mobile phases

1. Achiral Stationary Phases and Achiral Mobile Phases

Derivatization is required to separate diastereomers when we utilize an achiral stationary phase and an achiral mobile phase. Because of the inherent difficulty of preparation of derivatives and the attendant problems of by-products, this technique is not utilized to the extent that it should be. A few examples of stereoisomeric separations with a variety of derivatization reagents are shown in Table 7.

2. Achiral Stationary Phases and Chiral Stationary Phase Additives

As noted before, a variety of chiral eluent additives have been investigated. Some of these compounds are as follows:

- β-Cyclodextrin
- D-Galacturonic acid
- (R)-N-(3,5-Dinitrobenzoyl) phenylglycine
- N-(1R, 3R)-trans-Chrysanthemoyl-L-valine
- (+)-Tartaric acid
- (−)-Brucine

It may be recalled from the structure of β-cyclodextrin given in Figure 1 of Chapter 8 that it is a chiral toroidal-shaped molecule with a finite cavity formed by the connections of seven glucose units via 1,4-linkage. An example utilizing β-cyclodextrin for separation of derivatized amino acids is given here because cyclodextrins have been frequently used in separations of chiral

II EVOLVING METHODS AND METHOD SELECTION

TABLE 7 Stereoisomeric Separations by TLC

Compound	R_f (R, S)	Eluent	Derivatization reagent	Plate
Amphetamine	0.21, 0.14	Toluene/dichloromethane/tetrahydrofuran (5:1:1)	(S)-(+)-Benoxaprofen chloride	Silica gel 60
Methamphetamine	0.33, 0.27	As above	As above	As above
Methyl-benzylamine	0.28, 0.16	As above	As above	As above
Naproxen	0.53, 0.63	Chloroform/ethanol/acetic acid (9:1:0.5)	(1R, 2R)-(−)-1-(4-Nitrophenyl)-2-amino-1,3-propanediol	Silica gel F 254
Metoprolol	0.24, 0.28	Toluene/acetone (100:10)	(S)-(+)-Benoxaprofen chloride	Silica gel 60
Oxprenolol	0.32, 0.38	As above	As above	As above
Propranolol	0.32, 0.39	As above	As above	As above

compounds. Retention is related to the extent of the cavity of the cyclodextrin and additional interactions of the enantiomers with the oligomer. 2-Naphthylamide and *p*-nitroanilide derivatives of amino acids have been separated on Sil C18.50F plates with the following mobile phase containing β-cyclodextrin:[17]

- Aqueous solution (100 mL) containing sodium chloride (2.5 g)
- Urea (26 g)
- Acetonitrile (20 mL)
- 0.15 M β-cyclodextrin

The plate is developed to a distance of 7 cm (Fig. 9), and these racemates can be resolved:

- DL-Methionine-2-naphthylamide (spot #2)
- DL-Leucine-2-naphthylamide (spot m_1)
- DL-Leucine-*p*-nitroanilide (spot m_2)

The D and L alanine-*p*-nitroanilides, however, cannot be resolved by this method.

3. Chiral Stationary Phases and Achiral Mobile Phases

For enantiomeric separations, TLC plates can be coated with the following materials to provide useful stationary phases:

- Cellulose
- Chiral compounds
- Cyclodextrin
- Ligands

a. Cellulose

It may be recalled from Chapter 8 that cellulose is a linear macromolecule composed of optically active D-glucose units with helical cavities (Fig. 2,

FIGURE 9 Resolution of methionine and leucine.[17]

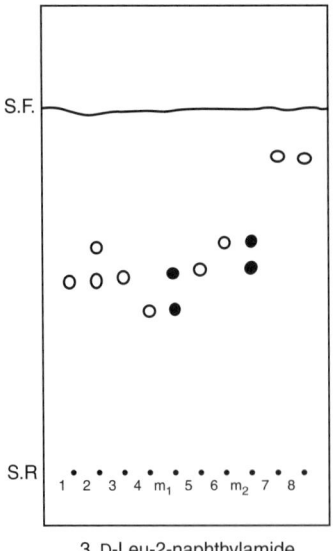

3. D-Leu-2-naphthylamide
4. L-Leu-2-naphthylamide
m_1-mixture of 3,4

Chapter 8). The separation of enantiomeric compounds is affected by the manner in which they fit into the lamellar chiral layer structure of the support. Microcrystalline triacetylcellulose plates are commercially available. These plates are stable with aqueous eluent systems as well as alcoholic eluents. They can be used with dilute acids and bases; however, they should not be used with glacial acetic acid and ketonic solvents.

b. Silica gel/chiral compound

A large number of amino acids can be resolved by TLC on silica gel/(+)-tartaric acid plates. For example, the PTH derivatives of amino acids have been resolved on silica gel/(+)-tartaric acid plates.[18] The mobile phase is composed of a mixture of chloroform/ethyl acetate/water (28:1:1), and the plates are developed for a distance of 10 cm. The hR_f values of various amino acids resolved is given in Table 8.

c. β-Cyclodextrin

A number of derivatized amino acids have been chromatographed as dansyl or naphthylamide derivatives on β-cyclodextrin plates with varying ratios of methanol and 1% triethylammonium acetate.[19] At pH 4.1, good resolution is generally achieved between D- and L-derivatives of the amino acid (Table 9).

d. Ligands

Using ligand exchange, aromatic amino acids have been separated on Merck RP-18 WF 254 S plates covered with copper acetate and LNDH (L-N-n-decylhistidine). The separation data are given in Table 10. The mobile phase consists of methanol/ACN/THF/water in the ratio of 7.3:5.9:33.9:52.9.[20]

II EVOLVING METHODS AND METHOD SELECTION 227

TABLE 8 hR_f Values of Resolved Enantiomers of PTH Amino Acids

DL Mixture	hR_f of D	hR_f of L
Methionine	16	83
Phenyl alanine	15	85
Valine	21	80
Tyrosine	16	95
Threonine	30	85
Alanine	12	55
Serine	10	84

Adapted from Reference 18.

TABLE 9 TLC of Amino Acids on β-Cyclodextrin Plates

Compound	R_f D	R_f L	Mobile Phase	Detection
Dansyl leucine	0.49	0.66	40/60	Fluorescence
Dansyl methionine	0.28	0.43	25/75	Fluorescence
Dansyl alanine	0.25	0.33	25/75	Fluorescence
Dansyl valine	0.31	0.42	25/75	Fluorescence
Alanine-β-naphthylamide	0.16	0.25	30/70	Ninhydrin
Methionine-β-naphthylamide	0.16	0.24	30/70	Ninhydrin

TABLE 10 Ligand Exchange

Amino Acid	Rm (L)	Rm (D)	α
Tryptophan	0.50	0.66	1.33
α-Methyltryptophan	0.56	0.73	1.38
5-Methyltryptophan	0.60	0.75	1.33
6-Methyltryptophan	0.63	0.72	1.19
Phenylalanine	0.07	0.25	1.28
α-Methylphenylalanine	0.18	0.42	1.43
Tyrosine	−0.03	0.13	1.21
α-Methyl tyrosine	0.06	0.24	1.27
DOPA [3-(3,4-dihydroxyphenyl) alanine]	0.08	0.24	1.24

Note: Adapted from Reference 20.

B. Gas Chromatography

As mentioned earlier, it is difficult to use gas chromatography for analysis of those compounds that are thermally unstable and are not volatile, because derivatization, along with its associated problems, is then required. The evolution of columns with improved thermal stability and gains in capillary column technology have created a renewal of interest in this technique. GC presents benefits such as high resolution and large peak capacity. In addition, it is effective for nonaromatic compounds employed in asymmetric synthesis, which are not readily separated by TLC or HPLC.

We should reexamine the problems encountered in carrying out GC of enantiomers:

1. The sample must be volatilized if it is not already volatile.
2. Racemization of the analyte may occur in chromatographic conditions.
3. Stationary phases can also be racemized at high temperatures.
4. Separations on a preparative scale are generally cumbersome.

Separation of enantiomers by GC has been achieved by the following two approaches:

- Resolution of prepared diastereoisomers on conventional columns
- Resolution of enantiomers on chiral stationary phases with or without derivatization of the analyte

I. Derivatization

Preparing appropriate derivatives for GC permits one to utilize conventional columns. Table 11 lists some derivatization reagents that can be used for alcohols, carboxylic acids (including amino acids and hydroxy acids), and amino compounds such as amines, amino acid esters, and amino alcohols.

Some of the disadvantages of derivatization are listed below.

- The chiral reagent must be enantiomerically pure.
- The subsequent diastereomer must be volatile and stable.
- It is possible that racemization will occur.

Volatility may be improved by the use of achiral reagents. These reagents may enhance stereoselective interactions. Also, racemization is avoided, along with the production of unwanted by-products. Table 12 shows derivatization reagents for a variety of compounds. It also shows the resultant products of interest.

TABLE II Derivatization Reagents

Compound	Derivatization reagent
Amino acids	(+)-3-Methyl-2-α-butanol
Carboxylic acids	(+)-3-Methyl-2-α-butanol
Hydroxy acids	(+)-3-Methyl-2-α-butanol
Amines	L-Chloroisovaleryl chloride
Amino alcohols	L-Chloroisovaleryl chloride
Amino acid esters	L-Chloroisovaleryl chloride
Alcohols	L-Chloroisovaleryl chloride

II EVOLVING METHODS AND METHOD SELECTION 229

TABLE 12 Derivatization Methods

Compound	Derivatization reagent	Product
Alcohols	Isocyanates	Urethanes
Carbohydrates	Phosgene	Carbonates
	Acylating reagents	Acylated products
Amines	Acylating reagents	Acylated products
Amino acids	As above	As above
Amino alcohols	Phosgene	Oxazolidinones
Hydroxy acids	Isocyanates	Urethanes, amides
Carboxy acids	Phosgene	1,3-Dioxolane-2,4-diones
Ketones	Hydroxylamine HCl	Oximes

2. Amino Acid Phases

Gil-Av and others[21] were the first to separate N-trifluoroacetyl (TFA) derivatives of amino acids by using N-trifluoroacetyl-L-isoleucine lauryl ester phase. They discovered that the D-enantiomers elute first. The separation was attributed to hydrogen bonding. In this respect, it may be informative to scrutinize significantly different α values and $-\Delta\Delta G$ values offered by the two selectands comprised of two different esters of N-TFA leucine (Table 13).

Pirkle, who designed a multidentate solute giving $\alpha = 121$ on silica gel to which N-β-naphthyl-L-alanine was bonded, introduced a real improvement. The need to obtain high α values especially for preparative separations might explain the popularity of HPLC for chiral compounds.

Other gas chromatographic phases of interest in this group are dipeptides and modified diamides, which have the following long-chain alkyl groups:

- *n*-Dodecyl-L-valine-t-butylamide
- *n*-Decanoyl-(S)-α-(1-naphthyl)ethylamide

An instance of a naturally occurring mode of hydrogen-bond association that may lead to separation of enantiomers has been reported by Feibush and others.[22] They studied the resolution of pharmaceutical compounds such as hydantoins, barbiturates, and glutaramides that have an –OC–NH–CO– moiety in the molecule. The authors linked N,N'-2,6-diaminopyridinediyl*bis*[S-2-phenylbutanamide] to silica gel by way of an *n*-undecyloxy handle. This CSP exhibited selectivity for the above-mentioned compounds. It can be demonstrated, on the basis of their studies, that the S selectand of the barbiturate fits better than the R isomer, as anticipated from the order of elution (Fig. 10). The two large substituents at the α-carbon of the amide group and the 1-position of the S-selectand (cyclohexyl) of the selector (phenyl) are *trans* to each other.

3. Polymeric Phases

A polysiloxane-based polymer with a chiral side chain is produced by copolymerization of dimethylsiloxane with (2-carboxypropyl) methoxy

TABLE 13 Resolution (α) for Different Selector, Selectand Systems

	Selector	Selectand	α (temp)	$-\Delta\Delta G^{**}$ (cal)
(I)	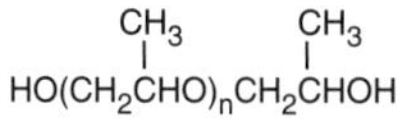 HO(CH₂CHO)ₙCH₂CHOH with CH₃ groups	Norbornanols	1.01 (1) (70 °C) GC	7
(II)	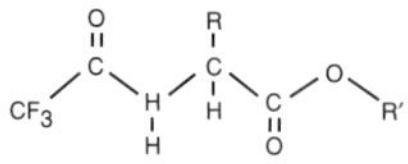 CF₃ acyl amino acid ester	N-TFA-Leu t-Bu Ester	1.08 (2) (90 °C) GC	55
(III)	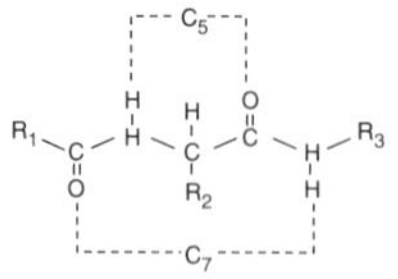 III: R₁ = n-C₁₁H₂₃; R₂: iPr, R₃ = t-Bu	N-TFA=Leu iPr Ester	1.34 (3) 130 °C [3.5–4.0] (extrapolated to 0 °C) GC	235 / 760
(IV)	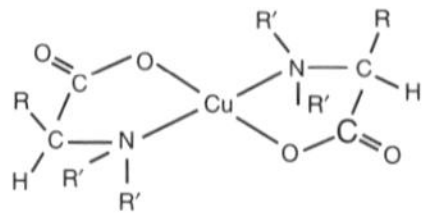 Cu complex	Valine	4.8 (4) (0 °C) HPLC	860
(V)	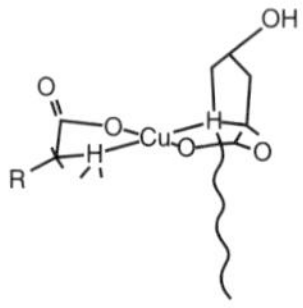 Cu-proline complex	Proline	10.9 (5) (20 °C) HPLC	1400
(VI)	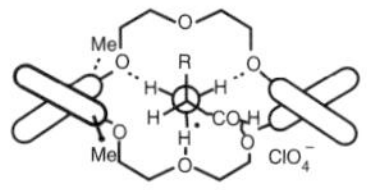 Crown ether / ClO₄⁻	[Phe]+/ClO₄⁻	30* (6) (0 °C)	1860
(VII)	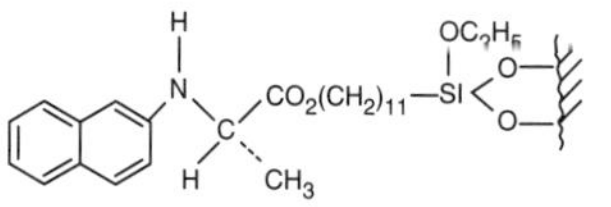 Naphthyl-amino silica CSP	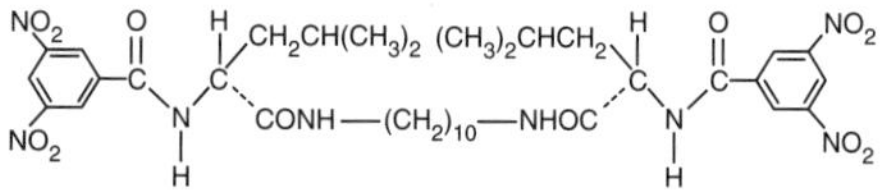 Dinitrobenzoyl bis-amide	121 (7) (25 °C) HPLC	2860

Note: Adapted from Reference 21.
*Enantiomer distribution constant between two immiscible solvents.
**$-\Delta\Delta G = RT \ln \alpha$

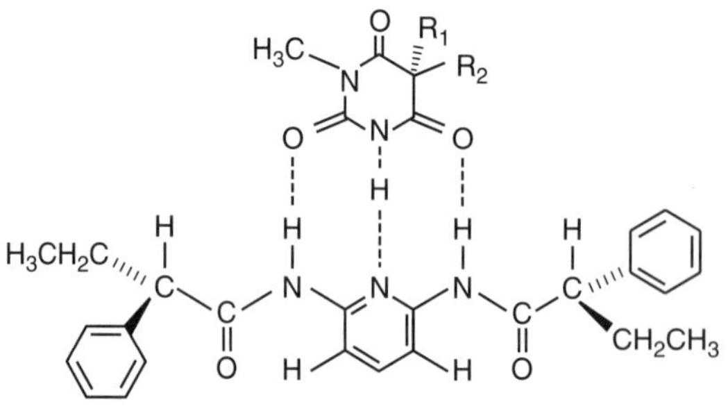

FIGURE 10 Interactions of solute and CSP.

silane and the coupling with L-valine-*t*-butylamide. This column can be used at temperatures up to 220 °C. It is useful for chromatographing amino alcohols and perfluoroacylated and esterified amino acids.

4. Cyclodextrins (CDs)

As mentioned before, these compounds are chiral, toroidal-shaped molecules that are composed of six to eight 1,4 linked glucose units:

- α-CD   6 units
- β-CD   7 units
- γ-CD   8 units

These compounds are crystalline, thermally stable, and insoluble in most organic solvents. Konig is credited with derivatizing hydroxyl groups of cyclodextrin by introducing alkyl or alkyl–acyl substituents to produce low-melting point-derivatized CDs. These CDs are stable and can be used for coating GC columns. Perpentylated β-CD, for example, can resolve TFA derivatives of sugars, amino acids, and amino alcohols.

The reactivities of 2-, 3-, and 6-hydroxyl groups of glucosidic units are different, so it is feasible to modify CDs with different substituents. For example, more polar permethyl-O-[(S)-2-hydroxypropyl]-CD offers enantio-selectivity that is opposite that of the analogous nonpolar alkyl derivatized CD. A number of commercially available CD phases are shown in the list below.

- 2,6-Di-O-pentyl-α-cyclodextrin
- 2,6-Di-O-pentyl-β-cyclodextrin
- 2,6-Di-O-pentyl-γ-cyclodextrin
- Permethyl-(S)-2-hydroxypropyl-α-cyclodextrin
- Permethyl-(S)-2-hydroxypropyl-β-cyclodextrin
- Permethyl-(S)-2-hydroxypropyl-γ-cyclodextrin
- 2,6-Di-O-pentyl-3-O-trifluoroacetyl-α-cyclodextrin
- 2,6-Di-O-pentyl-3-O-trifluoroacetyl-β-cyclodextrin
- 2,6-Di-O-pentyl-3-O-trifluoroacetyl-γ-cyclodextrin

5. Cyclodextrin Columns

A 30 m × 0.25-mm inner diameter β-cyclodextrin column with a film thickness of 0.25 μm was used in the resolution of ethosuximide. The run was operated isothermally at 160 °C, and a flame ionization detector was used.[23] The resolution was obtained in less than 9 minutes (Fig. 11). Separation of the enantiomers of hexobarbital and mephobarbital has been also reported on this column at 250/300 °C.[23] The resolution of both compounds was achieved in 17 minutes (Fig. 12).

Different functionalized cyclodextrins were utilized in the separation of variously substituted tetralins and indans, as shown in Table 14.[24] The column temperature ranged from 70 to 120 °C, and the alpha values ranged from 1.01 to 1.06.

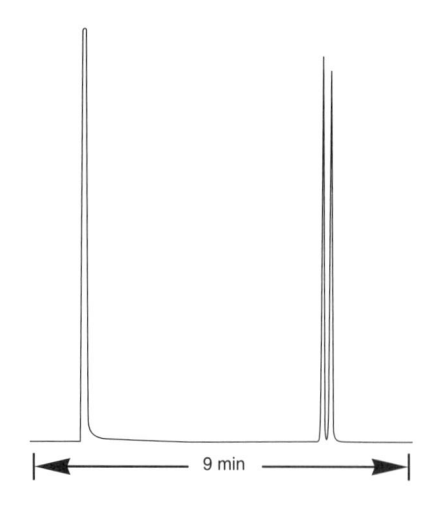

Column	: Cyclodex-B™
	30m × 0.25mm I.D.
	J&W PN 112-2532
Film Thickness	: 0.25 micron
Oven	: 160°C Isothermal
Carrier	: Hydrogen @ 1.10 mL/min (37 cm/s)
Injector	: Split 1:90:250°C
	1μL of 5μg/μL in methanol
Detector	: FID: 300°C
	Nitrogen make-up gas @ 30 mL/min

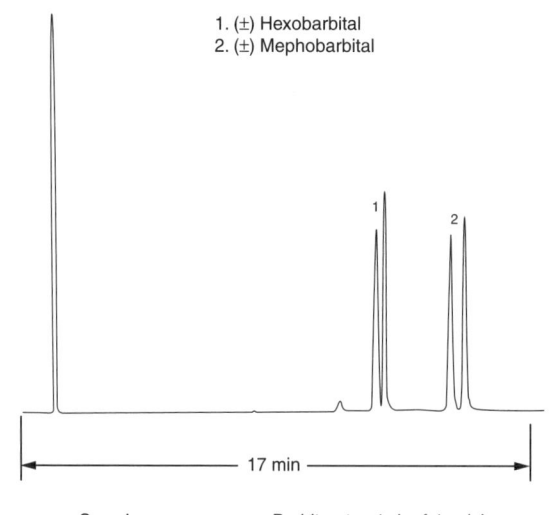

FIGURE 11 Resolution of ethosuximide by GC.[23]

1. (±) Hexobarbital
2. (±) Mephobarbital

Sample	: Barbiturates 1 μL of 4 μg/μL
	in methanol
Injection	: Split (75:1)
Oven temp	: 200°C
Injector/detector temp	: 250°/300°C
Carrier gas	: Hydrogen at 46 cm/s
Column	: J&W Cyciodex-B™
	30m × 0.25 mm × 0.25μ

FIGURE 12 Resolution of barbiturates by GC.[23]

TABLE 14 GC Separation of Hydroaromatic Biomarkers

Name	k'	α	Temp. (°C)	Column
Tetralin				
1,8-di-methyl-	6.9	1.03	110	10 m. B
2,7-di-methyl-	38.4	1.02	70	10 m. B
	21.1	1.05	90	10 m. A
1,5,8-tri-methyl-	10.3	1.05	120	10 m. B
2,6-di-methyl-	35.2	1.03	100	10 m. A
	12.42	1.04		
1,4-di-methyl-	21.5	1.06*	80	10 m. B
	58.4	1.01*	70	10 m. G
2-ethyl-	40.5	1.01	70	10 m. B
	12.9	1.02	100	10 m. A
Indan				
1-isopropyl-	19.4	1.02	90	10 m. A
1-propyl-	19.4	1.02	90	10 m. A
1-ethyl-	15.0	1.05	90	10 m. A

Note: Adapted from Reference 24
The k' value is for the first eluted enantiomer.
* This compound exists as a pair of enantiomers and a meso compound. This α value is for the enantiomeric pair only.
Key: G = 2,6-Di-O-pentyl-3-O-trifluoroacetyl-γ-cyclodextrin;
A & B = Permethyl derivatives of O-(s)(2-hydroxypropyl) cyclodextrin

6. Cross-Linked Columns

A large number of amino acids have been separated on a cross-linked CSP (polycyanoethylvinyl siloxane-L-valine-NHC(CH$_3$)$_3$) by temperature programming up to 190 °C (Fig. 13). The D enantiomers elute first.[25] Schurig[26] resolved racemic isomenthol, menthol, and neomenthol at 85 °C on 25 m × 0.25-mm inner diameter columns coated with *heptakis*(2,3,6-tri-O-methyl)-β-cyclodextrin on OV 1701.

C. High-Pressure Liquid Chromatography

HPLC methodology has been discussed at length in Chapter 10. Here, the discussion is limited to various ways that are available to resolve enantiomers by HPLC.

1. Chromatography of Diastereomeric Derivatives

Precolumn derivatization of an optically active solute is accomplished with another optically active molecule. This technique is useful for compounds that contain amino groups, hydroxyl groups, or carboxyl groups. Olefins, epoxides, and thiols can also be resolved. A number of examples of these separations are discussed in other sections of this book.

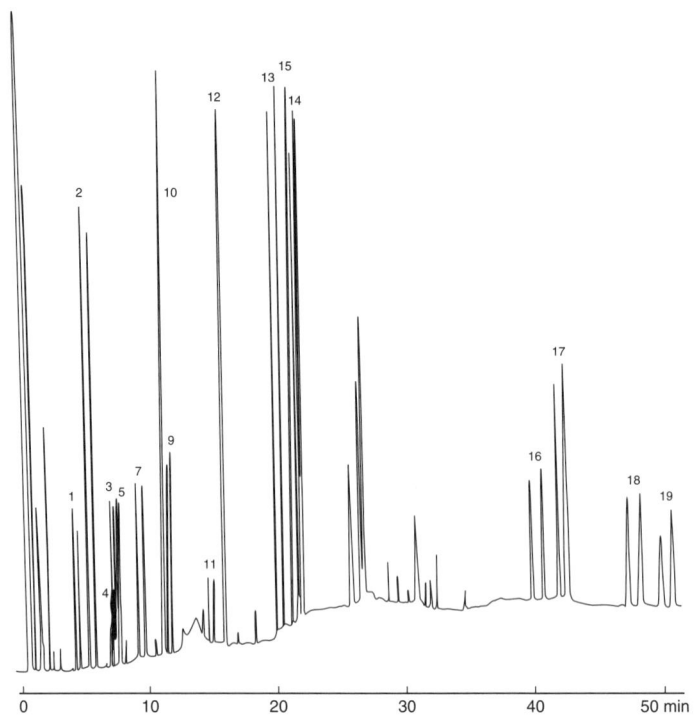

FIGURE 13 Enantiomeric separation of amino acids by GC.[25]

2. Resolution Using Chiral Mobile-Phase Additives

The resolution here depends on a diastereomeric complex being formed, with a chiral molecule added to the mobile phase. Chiral resolution occurs by virtue of the differences in the stabilities of diastereomeric complexes of various molecules, solvation in the mobile phase, or binding of the complexes to the solid support. The type of separation can be based on one of the following mechanisms:

- Transition metal complex
- Ion pairs
- Inclusion complexes

Examples of separations by these mechanisms may be found in Chapter 6 of Reference 2.

3. Resolution Using Chiral Stationary Phases

The enantiomers can be resolved by formation of diastereomeric complex between the solute and a chiral molecule that is bound to the stationary phase. Separation of enantiomeric compounds on CSP is due to the differences in energy between temporary diastereomeric complexes created between the solute isomer and the CSP; the bigger the difference, the greater the separation. The perceived retention and efficiency of a CSP, however, is established on the sum of all interactions between the solute isomer and the CSP, including achiral interactions.

II EVOLVING METHODS AND METHOD SELECTION

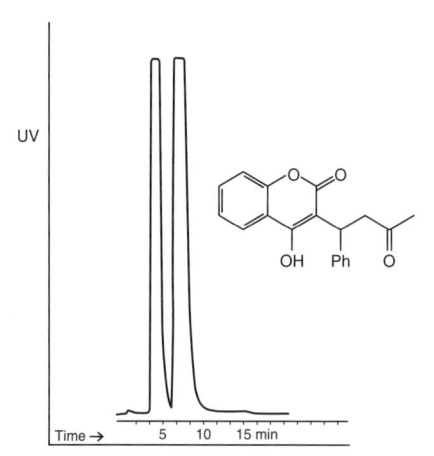

FIGURE 14 Preparative SFC of warfarin.[27]

D. Supercritical Fluid Chromatography

As mentioned before, the mobile phase in supercritical fluid chromatography (SFC) has a low viscosity and a high self-diffusion coefficient. This is attained by utilizing gases such as carbon dioxide, which at critical pressure and critical temperature render a supercritical fluid with low density. The density of carbon dioxide at 72.9 atm and 31.3 °C is 0.47 g/ml. The mobile phase can be additionally adjusted by addition of small amounts of polar modifiers. Therefore, SFC offers the advantages of both GC and HPLC in that the mobile phase is easily volatilized and that detectors used in GC can be utilized; the mobile phase selectivity of HPLC can still be utilized by working with various additives. The capacity to program the density or pressure is unique to SFC in that it permits enlarging the solvating power of the solvent by increasing the density.

Figure 14 illustrates preparative-scale separations for warfarin on Whelk-O 1 column.[27] The mobile phase consists of carbon dioxide containing 25% isopropanol and 0.5% acetic acid with UV detection at 260 nm.

E. Capillary Electrophoresis

Capillary electrophoresis (CE) has been established as quite a useful technique for resolution of a great number of compounds, including enantiomers. The principal advantage of CE is that it affords rapid, high-resolution separation of water-soluble components that are present in small amounts. The principles of electrically driven flow of ions in solution provide the basis of separation. Selectivity is achieved by adjustment of electrolyte properties such as pH, ionic strength, and electrolyte composition, or by the incorporation of electrolyte additives. Typical additives include organic solvents, surfactants, and complexing agents.

The intrinsic simplicity of CE has attracted many researchers to the utilization of the technique for chiral separations. The great potential of CE

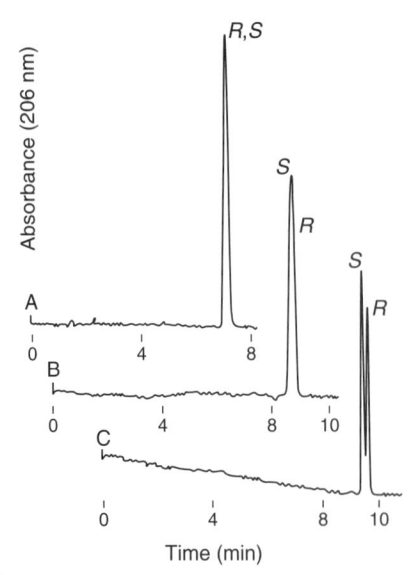

FIGURE 15 Electopherograms of propanolol.[28]

is sure to result in many useful applications in the coming years. Two methods of CE that are commonly used are:

- Free solution capillary electrophoresis (FSCE)
- Micellar electrokinetic capillary chromatography (MECC)

See Table 4 for a comparison of modes of separations used in CE with HPLC. The degree of chiral recognition attainable in chiral CE is often similar to that obtained with HPLC. However, the high efficiency offered by CE as compared to HPLC offers a significant advantage: it is possible to use it for analysis of small samples. Of course, this can be a limitation when a larger sample size is desired, as is frequently the case in preparative separations.

Propanolol has been resolved in 50 mM phosphate buffer (pH 2.5) containing 4 M urea, 40 mM β-cyclodextrin, and 30% methanol.[28] The effect of increasing concentrations of methanol can be seen in the electropherogram (Fig. 15). It has been demonstrated that the addition of 2% to 10% of maltodextrins [mixture of linear α-(1-4)-linked D-glucose polymers] can be useful for chiral resolution of nonsteroidal anti-inflammatory drugs such as flubiprofen, ibuprofen, and naproxen.[29]

Nishi[30] has employed bile salts to resolve a variety of chiral molecules, including diltiazem, trimethoquinol, and carboline derivatives. Generally, four bile salts (sodium cholate, sodium deoxycholate, sodium taurocholate, and sodium deoxytaurocholate) are used under neutral to alkaline conditions. The enantiomers of diltiazem, its related compounds, and trimethoquinol can be resolved only at pH 7.0 using sodium deoxytaurocholate. The method has been used to determine the optical purity of five batches of trimethoquinol down to 1% of the (R) isomer in the presence of (S) isomer (Fig. 16).

A number of recent handy publications dealing with a variety of separations may be found useful to the reader.[31-37] These publications cover

II EVOLVING METHODS AND METHOD SELECTION

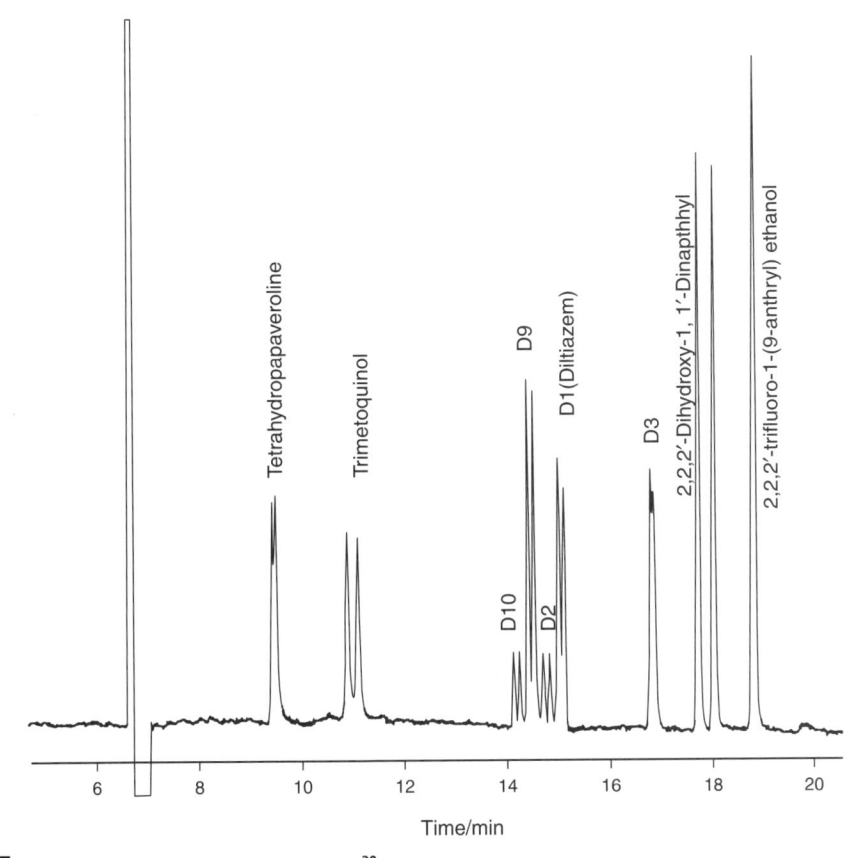

FIGURE 16 CE of trimethoquinol.[30]

method development strategies, the use of cyclodextrins, and macrocyclic antibiotics in chiral separations. For example, Lin[35] provides a useful list of separations of 63 drugs with seven cyclodextrins by CE.

F. Membrane Separations

Immobilized enzymes in hollow-fiber membrane reactors may be very useful for chiral separations.[2] In Figure 17, preparation of an L-acid is demonstrated. The hollow-fiber membrane laboratory-scale modules have the capacity to resolve 165 kg per year of a compound with a molecular weight of 150. With 20 full-scale modules, the capacity would be 100,000 kg per year. Membrane separations could conceivably provide the basis of chiral technology, which will be hard to match in terms of cost and convenience.

Further information on usefulness of membrane systems for enantioselective transport of chiral molecules may be found in Chapter 11 of Reference 2. Basic treatments of membrane theory, morphology, and transport considerations are included to better acquaint the reader with this mode of separation. Discussed in some detail are different types of enantioselective transport systems, grouped by structural class of the organic molecule. Details of separation

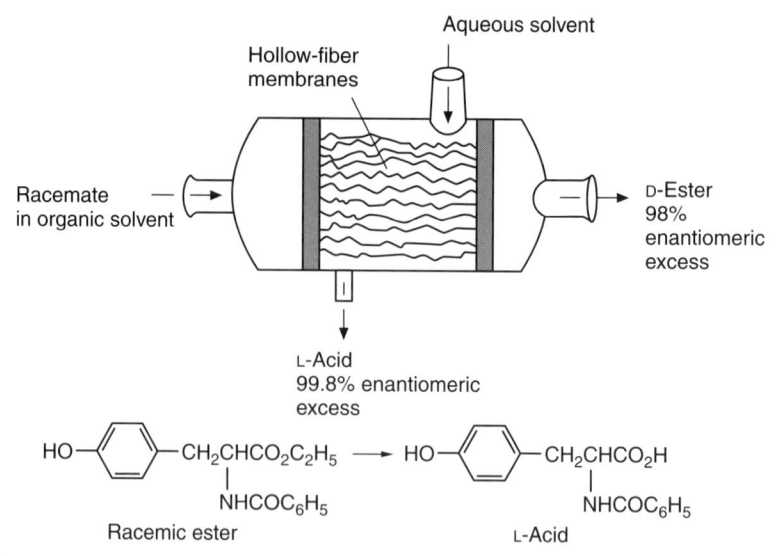

FIGURE 17 Preparation of an L-acid form from a racemic ester. (Courtesy of Sepracor.)

conditions have been provided for phenylpropanolamine, ephedrine, metazapine, phenylglycine, albuterol, terbutaline, propanol, and ibuprofen.

G. Enzymatic Methods

Enzymes may be classified into the following six classes:
- Hydrolases
- Isomerases
- Ligases
- Lyases
- Oxidoreductases
- Transferases

Enzyme-catalyzed reactions are ideally suited for production of single enantiomers because the interaction of proteins with small chiral molecules, in general, is highly stereospecific and shows clear preference for single enantiomers (see Chapter 4 in Reference 2). Hydrolases are by far the most important class of enzymes suitable for preparation of chiral compounds, and their applications have been developed extensively during the past two decades. Preparation of chiral alcohols, chiral acids and esters, amino acids, prochiral diols, and prochiral diesters are described.

When the chiral entity of interest is a carboxylic acid and the enzyme-catalyzed reaction involves making or hydrolyzing an ester or amide bond, one of the reaction components will be a free carboxylic acid. In these cases, the water solubility of the acid at basic pH provides an obvious and effective separation technique. Very different solubility characteristics may often be encountered for other functionalities. For example, the alcohol product and

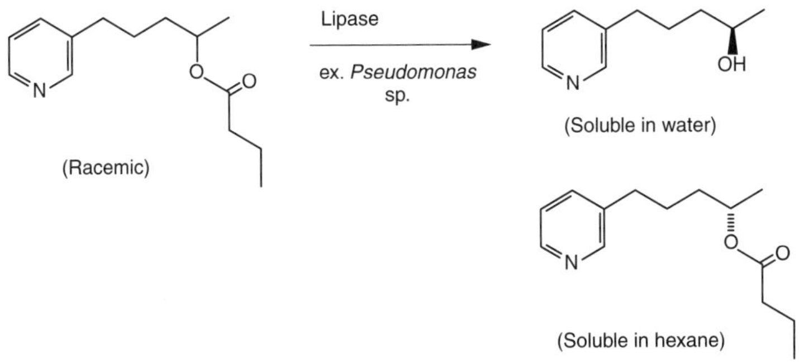

FIGURE 18 Enzymatic hydrolysis.

ester starting material of the enzymatic hydrolysis shown in Figure 18 can be completely separated by partitioning between water and hexane.[38]

The ketone and amine in the reaction mixture of aminotransferase-catalyzed amine reaction have been resolved by taking advantage of the dependence of their solubility on pH.[39] When selective extraction from aqueous medium is not feasible, fractional distillation or column chromatography can be used. Fractional distillation through a 30-cm Goodloe column is used to separate the C-13 ester and alcohol in the reaction mixture shown in Figure 19.

IV. COMPARISON OF GC, SFC, HPLC, AND CEC FOR A SELECTIVE SEPARATION

An impressive example of comparative separative powers of GC, SFC, HPLC, and CEC is described below for separation of enantiomers of hexobarbital by Schurig and coworkers.[40] It may be recalled that enantiomers have the same chemical structure, chemical mass, and physicochemical properties; they relate to each other as an object to its mirror image. Enantiomeric separations were evaluated on capillary Chiral–Dexcolumn (85 cm × 0.05 mm inner

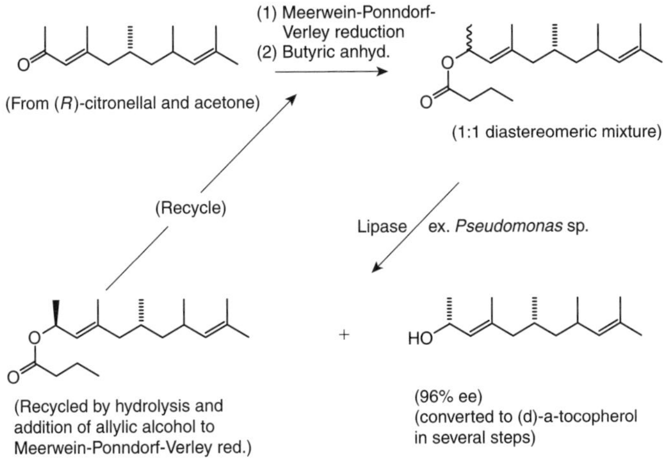

FIGURE 19 Enzymatic preparation of ester and alcohol.

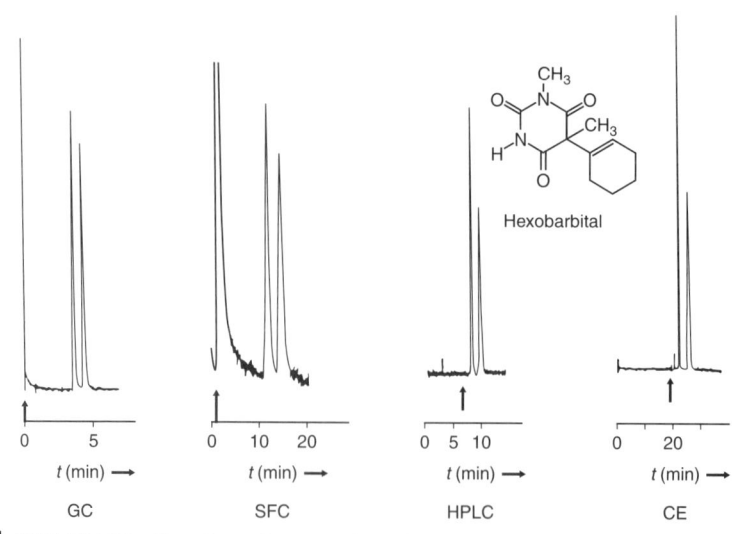

FIGURE 20 Enantiomeric separation of hexobarbital by GC, SFC, HPLC, and CEC.[40]

diameter with a film thickness of 0.15 µm). The results are shown in Figure 20. As a result of different diffusion coefficients in gases and liquids, analysis time and optimum efficiency in capillary HPLC in open tubular column and CEC are longer by four orders of magnitude as compared to GC. However, minimum plate height, optimum velocity, and total analysis time depend on the retention factor. The retention factor is large in GC, intermediate in SFC, and very small in HPLC and CEC for the first eluted enantiomer.

The following observations are of interest based on these separations:

- α for CEC and HPLC > SFC > GC
- R_S for CEC > HPLC > GC > SFC
- N (of the first peak) for CEC > HPLC = GC > SFC

It is noteworthy that CEC is the slowest method because of reduced electroosmotic flow in the coated capillary. Similar evaluations should be made prior to selection, optimization, and validation of any separation method.

REFERENCES

1. Karger, B. L., Snyder, L. R., and Horvath, C. *An Introduction to Separation Science*, Wiley, NY, 1973.
2. Ahuja, S. *Chiral Separations*, American Chemical Society, Washington DC, 1997.
3. Miller, J. M. *Chromatography: Concepts and Contrasts*, Wiley, NY, 1988.
4. Snyder, L. R., Glajch, J. L., and Kirkland, J. J. *Practical HPLC Method Development*, Wiley, NY, 1988.
5. Khaledi, M. G. *High Performance Capillary Electrophoresis*, Wiley, NY, 1998.
6. LC/GC, 5, p48
7. Giddings, J. C. *Anal. Chem.* 53:1170A (1981).
8. *Chromatography Abstracts*, Elsevier, New York, NY.
9. *GC and LC Literature: Abstracts and Index*, Preston, Niles, IL.
10. *Guide to GC Literature*, Plenum, New York, NY.

11. *A Guide to HPLC Literature*, Wiley, New York, NY.
12. *Compilation of GC Data*, Am. Soc. Testing Materials, Philadelphia, PA.
13. *Liquid Chromatographic Data Compilation*, Am. Soc. Testing Materials, Philadelphia, PA.
14. *Bibliography on Liquid Exclusion Chromatography*, Am. Soc. Testing Materials, Philadelphia, PA.
15. Scott, R. P. W. *Gas Chromatography* (A. Goldup, Ed.), Institute of Petroleum, London, p. 25, 1965.
16. Ahuja, S. *Selectivity and Detectability Optimizations in HPLC*, Wiley, New York, NY, 1989.
17. Lepri, L., Ceas, V., and Deideri, P. G. *J. Planar Chromatogr.* 4:338 (1991).
18. Bhusan, R. and Ali, I. *J. Chromatogr.* 392:460 (1987).
19. Alak, A. and Armstrong, D. W. *Anal. Chem.* 58:584 (1986).
20. Remelli *et al. Chromatographia*, 32:278 (1991).
21. Gil-Av, E., Feibush, B., and Charles, R. 6th International Symposium on GC and Related Techniques, Rome, September 20, 1966, through ACS Symposium series #471 (S. Ahuja, Ed.), Am. Chem. Soc. Washington, DC, 1991.
22. Feibush, B., Figueroa, A., Charles, R., Onan, K. D., Feibush, P., and Karger, B. L. *J. Am. Chem. Soc.* 108:3310 (1986).
23. *J & W Applications Booklet.*
24. Armstrong, D. W., Tang, Y., and Zukowski, J. *Anal. Chem.* 63:2858 (1991).
25. Lou, X., Liu, Y., and Zhou, L. *J. Chromatogr.* 552:153 (1991).
26. Schurig *et al., J.H.R.C.* 13:470 (1990).
27. Blum, A. M., Lynam, K. G., and Nicolas, E. C. *Chirality*, 6:302 (1994).
28. Fanali, S. *J. Chromatogr.* 545:437 (1991).
29. D'Hulst, A. and Verbeke, N. *J. Chromatogr.* 608:275 (1992).
30. Nishi, H., Fukuyama, T., and Terabe, S., *J. Chromatogr.* 515:233 (1990).
31. *Introduction to the Theory and Application of Chiral Capillary Electrophoresis*, Beckman Instruments, Fullerton, CA.
32. Rabel, S. and Stobaugh, J. *Pharm. Res.* 10:171 (1993).
33. Altria, K. D. *Capillary Electrophoresis Guidebook*, Humana, Totowa, NJ, 1995.
34. Heiger, D., Majors, R., and Lombardi, R. *LC.GC*, 15:14 (1997).
35. Lin, B., Zhu, X., Koppenhoefer, B., and Epperlein, U. *LC.GC*, 15:40 (1997).
36. Guttmann, A., Brunet, S., and Cooke, N. *LC.GC*, 14:32 (1996).
37. Ward, T. J. *LC.GC*, 14:886 (1996).
38. Manchand, P. S., Schwartz, A., Wolf, S., Bellica, P. S., Madan, P., Patel, P., and Saponsik, S. J. *Heterocycles*, 35:1351 (1993).
39. Coffen, D. L., Cohen, N., Pico, A. M., Schmid, R., Sebastian, M. J., and Wong, F. *Heterocycles*, 39:527 (1994).
40. Schurig, V., Mayer, S., Jung, M., Fluck, M., Jakubetz, H., Glauch, A., and Negura, S. In *The Impact of Stereochemistry on Drug Development and Use* (H. Y. Aboul-Enein and I. W. Wainer, Eds.), Wiley, New York, NY, 1997, p. 401.

QUESTIONS FOR REVIEW

1. List new evolving methods.
2. Describe briefly the advantages and limitations of the following methods:
   - Capillary electrophoresis
   - Supercritical fluid chromatography
   - Field flow fractionation
3. Why are separation methods for chiral compounds unique? What can we learn from them?

INDEX

Acid–base interactions, 58–59
Acidity, see pH
Activity coefficient (γ), 64
Adsorption
  in chromatographic methods, 9–10, 13, 43–44, 87, 89
  equilibria, 38, 43–44
Adsorption chromatography, 3, 12, 87–90, 221
  in HPLC, 179–180, 193–195
  on a liquid–solid column, 87–88, 89–90
  in TLC, 115, 123, 126
Adsorption isotherms, 43–44, 88
Alkanes, in TLC stationary phase, 116–117
Alkylnitrile bonded phase packings for HPLC, 195–196
Amine interactions in RPLC, 180, 181
Amino acids, resolution, 127, 225, 226, 229, 233
Amylase detection on paper chromatograms, 111
Artificial kidneys, 23–24
Ascending paper chromatography, 109

Ascending TLC, 124
Axial dispersion, 71
Azeotropic distillation, 26

Baclofen, RPLC retention mechanism, 180–182
Band broadening, 84
Barbiturates
  resolution, 231, 239–240
  RPLC retention, 174, 184
Bases, interactions with acids, 58–59
Basicity, see pH
Benzoic acid extraction, 27–28, 46
Bioautographic detection methods, 111
Boiling points, and intermolecular interactions, 52–53, 55, 56
Brownian motion, 69, 72
Bubble-cap columns, 25
Buchner funnels, 20

Capacity factor (k'), 47–48, 92, 95–96
  for GC, 146
  for HPLC, 160, 161, 165–167, 178

  in ion-pair mode, 199–200
  for RPLC, 172, 174, 176, 178
Capillary columns for GC, 135
Capillary electrochromatography, 216, 219–220
Capillary electrophoresis (CE), 216–219, 221, 222
  chiral separations, 235–237
  compared with HPLC, 219
Capillary gel electrophoresis (CGE), 218
Capillary isotachophoresis, 218
Capillary zone electrophoresis (CZE), 216–217
Carboxylic acids (chiral) separations by enzymatic methods, 238–239
Carrier gas for GC, 134–135, 136, 144, 170
Cartridges for TLC sample preparation, 121
Catecholamine derivatives, RPLC retention, 174
Cation exchange, 58–59
Cellulose
  papers, 20, 102–103

243

Cellulose (*Continued*)
  TLC plates, 118–119, 225–226
Centrifugation, 210
Chemical potential (μ), 38, 61, 64
Chemical potential profile (μ^*), 14
Chiral separations, 19, 117–119, 223–239
Chlorinated pesticides, GC separation, 84
Chlorophyll, column chromatography, 81–82
Cholesterol, TLC determination, 127
Chromatography, 1–2, 4, 8
  adsorption in, 9–10, 13, 43–44, 87, 89
  basic theory, 95–99
  classification, 9–14, 84–85, 87, 90
  comparison of methods, 93–95
    CE vs. HPLC, 219
    GC vs. HPLC, 155–157
    GLC vs. HPLC, 168–170
  diffusion in, 11, 70–71, 72–74, 98
  electron migration in, 9–10, 13
  history, 3–4, 81–82, 113–114
  and separation, 4–5
  speed, 83, 93, 95
    GLC, 136
  surface activity in, 9–10, 13
  see also individual methods
Chromosorb, 137–138
Clathration, 10
Cleanup of samples, 104, 120–122
Column chromatography, 85, 87–90, 154, 155
  history, 3, 81–82
  semiautomatic Craig process, 30–34, 82–83
  separation characteristics, 93
  for TLC sample preparation, 120, 121
  see also Gas chromatography (GC); Gas–liquid chromatography (GLC or GC); Gas–solid chromatography; HPLC; Packed columns
Column efficiency, 94, 95, 97
  for GC, 141
  for HPLC, 162–163, 178
Complexation, molecular, 58
Concentration of analytes, 103, 120
Cortisone, separation from urine, 121
Countercurrent extraction, 34–35
Craig multiple extraction process, 30–34, 82–83
Critical point, 41
Crystallization, 18, 19
Cyclodextrins
  in GC columns, 230–231
  as TLC eluent additive, 117, 224–225
  on TLC plates, 119, 225, 226

Darcy's law of fluid flow, 78
Debye interactions, 55–56, 141
Densitometric detection methods for paper chromatography, 111–112
Derivatization
  of cyclodextrins for GC columns, 231
  of enantiomers
    for GC, 228
    for TLC, 117, 224, 225, 226
  of GC samples, 155, 228–229
  of paper chromatography samples, 105–106
Desalination by reverse osmosis, 23
Descending paper chromatography, 109–110
Descending TLC, 124
Desorption, 73
Detectability, 16
  detection limits for GC, 143
Detection techniques
  for GC, 135, 142–145
  for HPLC, 156–157, 167
  for HPLC ion chromatography, 204
  for paper chromatography, 110–111
  for TLC, 125–126
Development of chromatograms
  for paper chromatography, 109–110
  for TLC, 124–125
Dextran
  for gel filtration chromatography, 86
  for TLC, 124
Dialysis, 5, 23–24, 122
Diatomaceous solid supports, 137–138
Differential column extraction systems, 34–35
Differential migration (dispersion), 84, 91
  in HPLC, 158
Diffusion, 11, 69–71, 72–74, 98, 99
  Fick's laws, 71–72
  in HPLC, 158–159
  in ion-exchange HPLC, 202
  rates, 74–77
Diffusion coefficient, 71, 72
  and diffusivity, 75–77
Diffusion constant, 75
Diffusion potential, 76–77
Digitalis glycosides, reversed-phase TLC, 130
Dimerization, in liquid–liquid extraction, 27
Dipole interactions, molecular, 54–56, 140–141
Dispersion forces
  intermolecular, 51
  and heat of vaporization, 52–53
  London forces, 50–52, 141
  and solubility parameter theory, 60–61, 65
Displacement chromatography, 91
Distillation, 12, 24–26, 221
  azeotropic/extractive, 26
  flash, 25
  fractional, 24–25
  steam, 25–26
  vacuum, 25
Distribution, 14
Distribution coefficient (K), 5, 26–27, 28–29, 95

INDEX

Distribution coefficient (*Continued*)
  at equilibrium concentration (K_x), 38, 65–67
  in distribution equilibria-based separations, 43, 44–45
  in dynamic equilibrium during column chromatography, 98
  in linear elution chromatography, 91, 92
  for mixtures, 29–30
  see also Partition coefficient
Distribution equilibria, 38
  effects of molecular interactions, 38, 65–67
  separations based on, 43–48
  and solubility parameter, 63–64
  of solutes, 44–45
  thermodynamics, 38–39, 65–67
Distribution isotherms, 43–45
Distribution ratio (D), 27–28, 45–47
Donor–acceptor interactions, 58, 60
Dragendorff reagent, 129–130
Dyes, TLC separation, 130–131

Eddy diffusion, in HPLC, 158
Efficiency of separation, 94, 95, 97
  by GC, 141
  by HPLC, 162–163, 178
Electrodialysis, 24
Electrolytes, diffusion rates in, 76–77
Electromigration, 9–10, 13
Electron capture detectors (ECD) for GC, 135, 143, 145
Electron donor–acceptor interactions, 58, 60
Electrophoresis, 11, 13, 211, 215–219
  see also Capillary electrophoresis (CE)
Electrostatic precipitation, 210
Eluent suppressor columns in ion chromatography, 204–205
Eluotropic series of solvents, Trappe's, 106–107
Elution chromatography, 91–93
Elutriation, 210
Enantiomers
  GC resolution, 227–233
  TLC resolution, 117–119, 224–227
Enclosed-bed chromatography, 87–90
Enzymes
  in catalytic chiral preparations, 238–239
  detection on paper chromatograms, 111
  immobilized, for chiral separations, 237
Equilibrium concentration distribution coefficient (K_x), 38, 65–67
Equilibrium processes in separations, 9–10, 13, 37–48
Equilibrium ultracentrifugation, 10
Ethanol extraction in paper chromatography, 104
Ethosuximide, GC resolution, 231
Eutectic point, 42
Evaporation, 17, 18, 24
Exclusion, 14
Extractant, 5
Extraction
  liquid–liquid, *see* Liquid–liquid extraction
  of mixtures, 29–30, 220–221
  of paper chromatography samples, 104
  see also Liquid–liquid partition chromatography
Extractive distillation, 26

Fick's laws of diffusion, 71–72
Filter papers, 20
  for paper chromatography, 102–103
Filtration, 17, 19–21, 209, 210
Flame ionization detectors (FID), for GC, 143, 144–145
Flash distillation, 25
Fluid flow fractionation (FFF), 213–215
Fluorescence measurement
  in paper chromatography, 105, 112
  in TLC, 125, 126
Forced diffusion, 70–71
Fraction capacity, chromatographic, 93, 94, 95
Fractional distillation, 24–25, 26
Frontal chromatography, 90

Gas chromatography (GC), 83–84, 85–86, 133–135, 211, 221, 222
  of amino acids, 229, 230, 233
  analyses, 145–146
  chiral separations, 227–233
  of chlorinated pesticides, 84
  columns, 135, 136–137, 231
  compared with HPLC, 155–157
  derivatization for, 155, 228–229
  detection techniques, 135, 142–145
  GC/MS instrumentation, 84
  quantitation, 146
  sample injection, 135
  separation characteristics, 93, 136
  selectivity, 84
  separation process, 135–136
  solid support, 137–138
  see also Gas–liquid chromatography (GLC or GC); Gas–solid chromatography (GSC)
Gas–liquid chromatography (GLC or GC), 12, 133–135
  compared with HPLC, 168–170
  history, 4, 26, 134
  mobile phase, 85
  stationary phase, 133, 134, 136, 138–140, 148–151
Gas–liquid separation methods, 9

Gas–liquid separation (*Continued*)
see also Gas–liquid chromatography (GLC or GC)
Gas–solid chromatography (GSC), 3, 5, 133–135
  stationary phase, 133, 134, 138
Gas–solid separation methods, 9–10
Gases
  diffusion rates in, 74–76
  separation by GSC, 138
Gel filtration, 4
Gel filtration chromatography, 86, 87
  see also Size-exclusion chromatography
Geometric mean rule for dispersion interactions, 52, 65
Gradient elution, for TLC, 125

Hard acids and bases, 58
Heat of mixing, 60–64
Heat of vaporization, 52–53
Height equivalent of theoretical plate, see HETP or HETU
Henry's law, 44
HETP or HETU (H), 35, 97–99
  for HPLC, 163
Hexobarbital, resolution, 231, 239–240
High-performance TLC (HPTLC), 114
High-pressure/high-performance liquid chromatography (HPLC or LC), see HPLC
Hollow-fiber membranes, 237
Homatropine methylbromide degradation products, TLC separation, 129–130
Homogeneous membranes, 22
Horizontal paper chromatography, 109, 110
Horizontal TLC, 124
HPLC, 12, 86, 153–154, 211, 221
  adsorption effects, 179–180
  applications, 155–156
  chiral separations, 233–234
  capacity factor (k'), 160, 161, 165–167
  column packings, 186–187, 195–196, 201–202, 204
  columns, 155
    evaluation, 182–184, 187–188
  compared with CE, 219
  compared with GC, 155–157
  compared with GLC, 168–170
  detection techniques, 156–157, 167
  efficiency, 162–163, 178
  history, 4, 26, 154–155
  instrumentation, 167
  mobile phase, 170–171, 185, 188–205
    additives, 192–193, 234
  resolution (R), 163–164, 165–167
  retention, 171–184, 194–195, 200–202
  retention parameters, 160–167
  retention volume, 161–162, 171
  reversed-phase, see Reversed-phase HPLC (RPLC)
  sample, 185–186
    injection, 157
  selectivity, 156, 168, 170–171, 178–179, 183–184
  separation
    characteristics, 93
    mechanism, 168–179
    process, 157–159
    strategy, 164–167, 222–223
  separation factor (α), 164, 165
  solvents, 189–192
  stationary phase, 156, 168–171, 177, 179–180, 234
  terminology, 4, 155
Hydrogen bonding
  in GC, 140, 229
  in paper chromatography, 107
  intermolecular, 56–57
Hydrogen bonding energy, in HPLC and RPLC, 175, 176
Hydrophobicity, in HPLC and RPLC, 174–175, 176, 197, 201–202

Ideal solutions, 60
Indans, GC resolution, 231
Induced dipoles, and molecular interactions, 54–55
Injection of samples
  for GC, 135
  for HPLC, 157
Interaction energy, molecular, see Molecular interaction energy
Interfaces, mass transfer through, 77–79, 158–159
Ion chromatography HPLC, 203–205
Ion exchange, 14, 58–59
  selectivity, 94, 202
Ion-exchange chromatography, 13, 86, 87, 181–182, 221, 222
  in HPLC, 200–203
  in TLC, 115, 124
Ion-exchange membranes, 22, 77
Ion-exchange resins, diffusion in, 77
Ion-pair chromatography
  in HPLC, 198–200
  in RPLC, 180–181

James, A.T., and Martin, 4, 134

Keesom forces, 54, 140
Kidneys, artificial, 23–24
Kovats polarity scale, 150–151
Kozeny–Carman equation for specific permeablilty, 79

Langmuir isotherm, 43–44
Ligand-exchange HPLC, 201

INDEX

Ligand-exchange TLC plates, 119, 226
Linear elution chromatography, 91–93
Liquid chromatography (LC), 85, 154, 221
  see also HPLC; Liquid–liquid partition chromatography
Liquid–liquid extraction, 4–5, 10, 12, 26–30, 95
  countercurrent, 34–35
  of mixtures, 29–30
  successive (multiple), 28–30
    semiautomatic Craig process, 30–34, 82–83
Liquid–liquid partition chromatography, 4–5, 12, 87, 95
  history, 3, 26
  similarity to countercurrent extraction, 34–35
  TLC, 115, 124
  see also Paper chromatography
Liquid–solid adsorption chromatography (LSC), 87–88, 89–90, 193–195
Liquids
  diffusion rates in, 74, 76–77
  separation methods, 10
Load capacity, chromatographic, 93, 94, 95
London forces, 50–52, 141
Longitudinal diffusion in HPLC, 159

McReynolds polarity scale, 151
Martin, A.J.P., 175
  and James, 4, 134
  and Synge, 3–4, 95, 134
Martin equation for distribution coefficients, 66–67
Masking agents, 47
Mass spectrometry, 11
Mass transfer
  in HPLC, 158–159
  through phase interfaces, 77–79

Mass transport in separation processes, 14
Membrane filters, 20, 103
Membranes
  for chiral separations, 237
  thin, 22–23
Methanol, 56, 57
Micellar electrokinetic capillary chromatography (MECC), 218
Microbiological detection methods for paper chromatograms, 111
Microporous membranes, 22
Mixtures
  separation methods, see Separation methods
  of solutions, solubility parameter theory, 60–65
Mobile phase, 5, 82
  for adsorption chromatography, 89–90, 193–195
  flow and diffusion in, 73–74
  for GC, 85
  for HPLC, 170–171, 185, 188–205
  for paper chromatography, 106–107, 108–109
  for RPLC, 176–177, 196–198
  for SFC, 212
  for TLC, 114–115, 116, 123–124
    additives, 117–118
    of enantiomeric compounds, 117, 119
Molecular diffusion, 70, 71, 73
Molecular interaction energy, 50
  due to acid–base interactions, 59
  due to dipole interactions, 54–56
  due to hydrogen bonding, 56–57
  due to London forces, 50–52
  and solubility parameter theory, 63–64
  total, 59–60, 65
Molecular interactions, 49–67
  dispersion forces, 50–53

effect on separation equilibria, 38, 49–50, 59–60
  in GC, 140–141, 156
  in HPLC, 156
  hydrophobic bond formation, 174–175
  nondispersion forces, 53–59
  in RPLC, 171, 174–175
  and solubility parameter theory, 60–61, 63–64, 65
  solvophobic, 171
  specific, 54, 141
Molecular probes for HPLC column evaluation, 182–184
Multiple development of TLC plates, 125
Multiple liquid–liquid extraction, 28–34, 82–83

Natural gas, GC analysis, 83
Nernst distribution law, 45
Nondispersion forces
  intermolecular, 53–59
  Debye, 55–56, 141
  hydrogen bonding, 56–57
Nonlinear elution chromatography, 91
Normal-phase HPLC, see HPLC
Normal-phase TLC, 116

Octanol/water partition, and RPLC retention, 172–175
Open-bed chromatography, 90
Osmosis, reverse, 23

Packed columns
  flow through, 78–79
  for GC, 137
  for HPLC, 154, 155, 186–187, 195–196, 201–202, 204
  van Deemter equation for H, 141–142
Paper chromatography, 3, 83, 85, 101–112, 222

Paper chromatography (*Continued*)
  chromatogram development, 109–110
  component detection, 110–111
  mobile phase, 106–107, 108–109
  papers, 102–103
  quantitation, 111–112
  sample preparation, cleanup, and derivatization, 103–106
  separation characteristics, 93
  stationary phase, 107–108
Particulate samples, separation, 210–211
Partition, 14
Partition chromatography, *see* Liquid–liquid partition chromatography
Partition coefficient
  in GC, 141
  octanol/water, and RPLC retention, 172–175
  in RPLC, 172
  see also Distribution coefficient (*K*)
Peak capacity, chromatographic, 94
Peak leading and tailing, 88
Permeable barriers, 11, 13, 23–24
  semipermeable membranes, 21–23, 202
Pesticides
  detection on paper chromatograms, 111
  GC separation, 84
pH
  acidity and hydrogen bonding, 57
  and ionization, 28, 46–47, 87
  and precipitation, 19
Phase equilibria, 37, 40–42
Phase rule, 40–42
Phases, 37
Phenylbutazone degradation products, TLC separation, 127
Physicochemical parameters, and retention in HPLC and RPLC, 175–178
Planar chromatography, 90

see also Paper chromatography; Thin-layer chromatography (TLC)
Polar molecules, interactions with nonpolar phases, 54–56
Polarity
  in GLC
    liquid phases, 150–151
    samples, 147–148
    stationary phase, 139–140, 150–151
  in HPLC and RPLC, 176, 178, 189–192
  in TLC solvents, 123
Polarity index (*P'*), 189–192
Polyamide stationary phase for TLC, 116, 124
Precipitation, 10, 18–19, 210
Prednisone, TLC assay, 127
Pressure diffusion, 70
Propranolol, CE resolution, 236
Pumps for HPLC, 167

Radial paper chromatography, 109, 110
Radioactive samples, detection, 112, 126
Ramsey, W., 3, 134
Rate processes in separations, 10–11, 70, 72–74, 78–79
Rates of diffusion, 74–77
Reflux, 24
Resolution (*R*$_s$), 99
  for descending paper chromatography, 110
  for GC, 146
  for high-pressure TLC, 114
  for HPLC, 163–164, 165–167
Retention
  chromatographic modes, 87, 89
  in HPLC and RPLC, 171–184, 194–195, 197–198, 200–202
Retention parameters
  retention *R*, 92–93
  retention time *t*$_R$, 92, 96–97
  for GC, 145–146

  for HPLC, 160–161, 163–164
  retention volume *V*$_R$, 91–92, 97
    for HPLC, 161–162, 171
    for RPLC, 175–176
  R$_{forces}$, 93
    for paper chromatography, 110
    for TLC, 124, 125
  R$_M$, 93
  see also Capacity factor (*k'*)
Reverse osmosis, 23
Reversed-phase HPLC (RPLC)
  adsorption effects, 179, 180, 222
  of baclofen, 180–181
  capacity factor, 172, 174, 176
  column packings, 196
  mobile phase, 176–177, 196–198
  retention
    mechanism, 171–172, 174–175, 180–181, 197–198
    and physicochemical effects, 175–178
    prediction, 183–184
    retention parameters and octanol/water partition, 172–175
  selectivity, 177, 178–179, 183–184, 197, 198
  stationary phase, 179, 180
Reversed-phase paper chromatography, 108
Reversed-phase TLC, 116–117, 124
  of digitalis glycosides, 130
Rohrschneider polarity scale, 123, 150–151

Samples
  HPLC
    recovery, 157
    selection, 185–186
  methods of introduction, 90–93
  nonparticulate, 211
  paper chromatography, preparation, 103–104

INDEX

Samples (*Continued*)
  particulate, 210–211
  TLC, preparation, 119–122
  types, and separation methods, 93, 94, 220, 222
Schleicher and Schuell paper grades, 102–103
Sedimentation, 209, 210
Selectivity, 14–15, 93, 94–95
  in HPLC, 156, 168, 170–171, 177, 183–184, 189–192
  in ion-exchange HPLC, 202
  in liquid–solid chromatography, 193–194
  in RPLC, 177, 178–179, 183–184, 192, 197, 198
Semiautomatic multiple liquid–liquid extraction, 30–34, 82–83
Semipermeable membranes, 21–23, 202
Separation, thermodynamics, 38–39
Separation factor (α), 48, 94, 97
  for GC, 146
  for HPLC, 164, 165, 167
Separation methods, 1, 2–3, 209–212
  chiral, 19, 117–119, 223–239
  and chromatography, 4–5
  classification, 7–13, 14–15
  equilibrium processes, 9–10, 13, 37–48
  effect of molecular interactions, 38, 59–60
  examples in everyday life, 6–7
  mechanical, 7–8, 210
  rate processes, 10–11, 70, 72–74, 78–79
  selection strategies, 220–223
  for HPLC, 164–167, 222–223
  unified, 14–15
  see also individual methods
Silica, reversed-phase (RP silica), 180
Silica gel
  for GSC gas separation, 138
  in HPLC column packings, 186–187, 195
Silica gel/(+)tartaric acid TLC plates, 119, 226
Size-exclusion chromatography (SEC), 10, 86, 87, 221, 222
  by TLC, 116
Slab electrophoresis, 216
Snyder solvent selectivity triangle, 183–184, 189–192
Soft acids and bases, 58
Solids
  diffusion rates in, 77
  gas–solid separation methods, 9–10
Solubility parameter (δ), 63–64
Solubility parameter theory, 60–65
  and retention in HPLC, 178–179
Solutions
  diffusion rates in, 76–77
  influence of dipole interactions, 54–55
Solvent precipitation, 19
Solvent strength in HPLC, 165, 170, 189–192, 193–194, 198
Solvents
  in mobile phase, *see* Mobile phase
  for sample preparation, 104, 119
  Snyder selectivity triangle, 183–184, 189–192
Solvophobic interactions in RPLC, 171, 197
Sorption chromatography, 73, 87–90
  see also Adsorption chromatography; Liquid–liquid partition chromatography
Specific molecular interactions, 54, 141
Speed
  of chromatographic analysis, 83, 93, 95
  GC, 136
  travel rate of sample in mobile phase, 96, 97–98
Stationary phase, 5, 82
  for GLC, 133, 134, 136, 138–140, 148–151
  for GSC, 133, 134, 138
  for HPLC, 156, 168–171, 177, 179–180, 234
  for paper chromatography, 107–108
  for RPLC, 180
  for TLC, 114–117, 118–119, 224–227
Steam distillation, 25–26
Steric exclusion chromatography, 4
Steroids, HPLC separation, 183, 184
Subcritical fluid chromatography (subFC), 213
Sublimation, 9, 24
Substituent effects on retention in HPLC and RPLC, 175–176
Successive (multiple) liquid–liquid extraction, 28–34, 82–83
Sulfonamides, TLC separation, 127
Supercritical fluid, 41
Supercritical fluid chromatography (SFC), 86, 212–213
  chiral separations, 235
  separation characteristics, 93
Suppressor ion chromatography, 204–205
Surface activity, 9–10, 13, 88, 89
Surface filters, 21
Synge, B.L.M., 3–4, 95, 134

Temperature effects
  on GC, 143, 150
  on HPLC, 156, 177, 202–203
  on TLC, 125
Tetralins, GC resolution, 231
Theoretical plates, 34, 35, 97
  height equivalent (HETP), 35, 97–99, 163
  total number (N), 97
  GC requirement, 146
  in HPLC, 162, 165, 166
Thermal conductivity detectors (TCD), for GC, 135, 142, 143–144
Thermal diffusion, 70

Thermal fluid flow
    fractionation, 214
Thermodynamics
  of distribution equilibria,
      38–39
  solubility parameter
      theory, 60–65
Thin-layer chromatography
    (TLC), 83, 85,
    113–114, 211,
    221–222
  of amino acids, 127, 225,
      226
  by adsorption, 115, 123,
      126
  by ion exchange, 115, 124
  by partition, 115, 124
  by size exclusion, 116
  chiral separations,
      117–119, 224–227
  chromatogram develop-
      ment, 124–125
  detection techniques,
      125–126
  on dextran gels, 124
  of dyes, 130–131
  high-performance, 114
  mobile phase, 114–115,
      123–124, 224–227
  normal-phase, 116
  of pharmaceuticals,
      127–130
  on polyamides, 116, 124
  preparative, 114
  sample preparation, 122

  quantitation, 126
  reversed-phase, 116–117,
      124, 130
  separation characteristics,
      93
  stationary phase, 114–117,
      118–119, 224–227
  of vitamins, 130
Tie line, 41
Tiselius, Arne, 216
Trace analysis, 7, 119–120
Trappe's eluotropic series of
    solvents, 106–107
Travel rate of sample in
    mobile phase, 96,
    97–98
Triacetylcellulose TLC plates,
    119
Triple point, 41
Tswett, M.S., 3, 4, 81, 85, 134
Two-directional development
  of paper chromatograms,
      110
  of TLC plates, 125

Ultracentrifugation, 210, 222
Ultrafiltration, 23, 222
Ultratrace analysis, 7,
    119–120
Urine, removal from TLC
    samples, 121
UV absorption
    measurement, 105,
    125, 126

Vacuum distillation, 25
van Deemter equation,
    98–99, 141–142
van der Waals forces, 140
van der Waals molecular
    separation, 50
van der Waals volume, and
    HPLC retention, 175,
    176
Vapor pressure, and GLC
    separation, 85–86
Vaporization, see
    Evaporation; Heat of
    vaporization
Vitamins, TLC separation,
    130

Warfarin, preparative-scale
    SFC, 235
Water
  hydrogen bonding in, 56,
      57
  miscibility with
      paper chromatog-
      raphy mobile phase,
      107
Whatman paper grades, 20,
    102–103
Wilke–Chang equation for
    diffusivity, 76

Zone refining, 12